CONCEPTUAL FRAMEWORKS IN GEOGRAPHY
GENERAL EDITOR: W. E. MARSDEN

Soils, Vegeta

Greg O'Hare B.Sc. M

Department of Geography,
Derbyshire College of Higher Educat

Maps and diagrams drawn by Ann R

Oliver & Boyd

Acknowledgements

The author and publishers wish to thank all those who gave their permission to reproduce copyright material in this book. Information regarding sources is given in the captions. Photographs provided by Joy Tivy were donated by the Department of Agronomy, Cornell University, USA.

Cover illustration by Kate Isles

To my wife, Lyn

Oliver & Boyd
Addison Wesley Longman Limited
Edinburgh Gate
Harlow
Essex CM20 2JE
An Imprint of Longman Group UK Ltd

ISBN 0 05 004237 8

First published 1988
Tenth impression 1997

Set in 10 on 12pt Linotron Times Roman

Produced by Longman Singapore Publishers Pte Ltd
Printed in Singapore

The Publisher's policy is to use paper manufactured from sustainable forests.

Contents

4

Editor's Note

An encouraging feature in geographical education in recent years has been the convergence taking place of curriculum thinking and thinking at the academic frontiers of the subject. In both, stress has been laid on the necessity for conceptual approaches and the use of information as a means to an end rather than as an end in itself.

The central purpose of this series is to bear witness to this convergence. In each text the *key ideas* are identified, chapter by chapter. These ideas are in the form of propositions which, with their component concepts and the inter-relations between them, make up the conceptual frameworks of the subject. The key ideas provide criteria for selecting content for the teacher, and in cognitive terms help the student to retain what is important in each unit. Most of the key ideas are linked with assignments, designed to elicit evidence of achievement of basic understanding and ability to apply this understanding in new circumstances through engaging in problem-solving exercises.

While the series is not specifically geared to any particular 'A' level examination syllabus, indeed it is intended for use in geography courses in universities, polytechnics and in colleges of higher education as well as in the sixth form, it is intended to go some way towards meeting the needs of those students preparing for the more radical advanced geography syllabuses.

It is hoped that the texts contain the academic rigour to stretch the most able of such candidates, but at the same time provide a clear enough exposition of the basic ideas to provide intellectual stimulus and social and/or cultural relevance for those who will not be going on to study geography in higher education. To this end, a larger selection of assignments and readings is provided than perhaps could be used profitably by all students. The teacher is the best person to choose those which most nearly meet his or her students' needs.

W. E. Marsden
University of Liverpool.

Preface

This text provides an up-to-date analysis of soils and vegetation and of the ecosystems, or general environments, in which they occur. It has heen written in response to the increasing interest in these subjects in recent years and the resulting expansion of syllabuses and examination questions, especially at 'A' level.

The book follows a logical, if not a didactic, format. The elements and properties of soils are introduced in Chapter 1. Chapter 2 examines a range of British soil types, covering their nature, formation and distribution. Chapter 3 turns to vegetation and focuses on its character, distribution and field survey. The information in Chapters 1 to 3 serves as a foundation for the subject of ecosystems, which is introduced in Chapter 4. Chapters 5 and 6 cover the structure and function respectively of the main global ecosystems (biomes) and the climates, soil, vegetation and animals, including humans, that they contain. In Chapter 7, elements from previous chapters are combined in an investigation of those ecosystems which have been particularly disturbed by human agency. This serves to remind us of the fragility and inter-connectedness of the global ecosystem. Then, in order to link theory more with practice and in view of the importance of project work in biogeographical syllabuses, two chapters of a practical nature complete the book. Chapter 8 deals with the laboratory analysis of soils and their properties, while Chapter 9 concerns the field observation of soil profiles, complementing that for vegetation in Chapter 3.

Many of the ideas set out in the book have been gleaned from a wide variety of sources, not least from advanced, classical, ecological and biogeographical texts such as Whittaker's *Communities and Ecosystems* (1975) and Tivy's *Biogeography* (1982). Wherever possible, ecological concepts have been reinforced by using recent quantitative data sources but the use of technical terminology has been kept to a minimum. To assist understanding, therefore, a glossary of ecological terms is provided at the end of the text.

Finally, I would like to thank all those who have helped in the preparation of the book.

Greg O'Hare

1 *The Nature and Properties of Soils*

A. Introduction

1. What is soil?

Soils constitute the uppermost layer of the earth's crust or *mantle*. The rocks on the earth's surface are ground down and weathered into smaller fragments by wind, rain and sun. The weathered mantle, or *regolith*, is further altered at the surface by the addition of the rotten remains of plants and animals. When organic matter is incorporated within the regolith in this way, soils begin to form. Indeed, soils have been defined as 'the biologically modified weathered mantle'. Another, simpler definition of soil is 'the stuff in which plants grow'. This definition again emphasises the biological importance of the soil.

2. The soil profile

Figure 1.1 Soil profile showing A-, B- and C-horizons

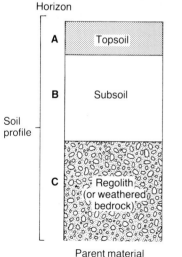

Individual soils are identified and described chiefly by means of a *soil profile*. As shown in Figure 1.1, and Plate 2.1 on page 73, a soil profile is a vertical column of soil. It may be seen, for example, in the exposed face of a pit or cutting. The profile usually consists of layers of differing material called *horizons*. Each horizon is distinct in colour, stoniness, moisture content, plant remains and so on. The different horizons run roughly parallel to the ground surface and lie above bedrock. This bedrock may represent the *parent material*, i.e. the principal source of weathered rocks from which the soil has formed. However, very often the uppermost horizons of soils are found in a different and more recent parent material, such as glacial till or wind-blown loess. In such cases, the soils reflect the underlying surface deposits and not the solid geological formations.

The letters A, B, C are commonly used to designate the main soil horizons. The uppermost horizon, or *topsoil*, is called the A-horizon. This is a layer in which material is both added and lost. The B-horizon, or *subsoil*, is principally a zone where material accumulates. The lower C-horizon is the weathered mantle, or regolith. Chapter 9, section B (page 190) lists the labels used to describe soil horizons. This information may be referred to throughout the book.

3. The soil in three dimensions

Because the soil extends laterally in all directions over the surface of the

Figure 1.2 Relationship between soil horizon, soil profile, soil pedon, soil polypedon and soil type (Source: Birkeland, 1984)

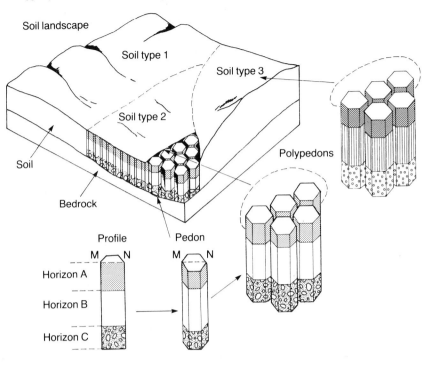

Figure 1.3 Four basic constituents of soils. This example illustrates the average proportional arrangement of soil constituents in a typical agricultural soil in the UK.

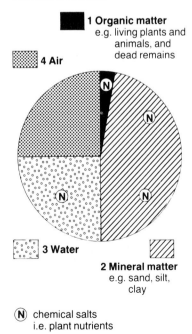

1 Organic matter
e.g. living plants and animals, and dead remains

4 Air

3 Water

2 Mineral matter
e.g. sand, silt, clay

Ⓝ chemical salts
i.e. plant nutrients

earth, it is probably more realistic to consider individual soils as three-dimensional bodies, rather than as two-dimensional soil profiles. The three-dimensional column of soil shown in Figure 1.2, with a surface area of about 10 m², is the smallest volume that can be called 'a soil'. This unit is termed a soil *pedon*. Similar pedons when grouped together form *polypedons* and eventually individual *soil types*. The different soil types seldom have sharp boundaries between them unless they are determined by a marked change in geology or some other environmental factor. Rather they tend to merge gradually with each other over the land surface. Collectively they constitute the *soil landscape*.

4. Soil materials and their properties

All soils are composed of the same basic materials, although the proportions of these vary greatly from one soil type to another. Individual soils are made up of four main constituents: air, water, mineral particles and organic (plant and animal) material. Figure 1.3 shows the *average* proportional arrangement of the soil constituents in a typical agricultural soil in the UK. One-half of the soil (by volume) is not solid at all, being made up of air and water in roughly equal parts. Most of the solid portion is mineral matter, such as sand and clay, with trace amounts of chemical salts. A fairly small amount, 5% or less, is organic material.

The four principal soil constituents shown in Figure 1.3 are not present as separate components, however. They are usually intimately mixed with

8

each other. This close intermixture gives rise to a number of important properties including *soil texture*, *soil structure* and above all *soil fertility*. These soil properties, together with the nature and properties of the soil constituents themselves, are the subject of the remaining sections of this chapter.

ASSIGNMENTS
1. (a) *Suggest why a soil is more than simply weathered rocks.*
   (b) *Distinguish between the following: parent material, regolith, soil profile, soil horizon.*
2. *Using Figures 1.1 and 1.2, show how soils can be described as two-dimensional and three-dimensional objects.*
3. (a) *Using Figure 1.3, describe the chief constituents of soils.*
   (b) *Comment on the advantages and disadvantages of this diagram as a model of actual soil composition.*

B. The Mineral Particles

1. The products of weathering

The mineral matter of a soil forms its 'skeleton' or physical framework. The individual mineral particles of a soil are formed by the weathering of the parent rock. Parent rocks, such as granite, consist of two types of mineral: hard minerals, such as quartz, and softer minerals, such as mica and feldspar. The hard minerals weather to give chemically resistant remains of sand and silt (i.e. primary minerals), while the softer minerals weather to form chemically altered products of clay (i.e. secondary minerals) and traces of mineral or chemical salts (see Figure 1.4).

(a) Sand and silt

The sand and silt found in soils are composed of grains of quartz (see

Figure 1.4   Principal sub-divisions of the mineral matter of a soil

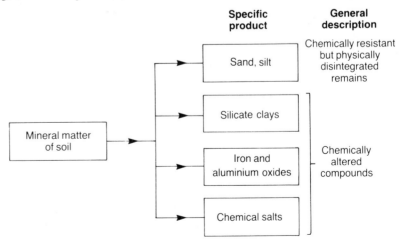

Sand grains
(mostly quartz)

Diameter
0.05–2.00 mm

Silt grains
(mostly quartz)

Diameter
0.002–0.05 mm

Clay particles
(iron oxides
and silicates)

Diameter
<0.002 mm

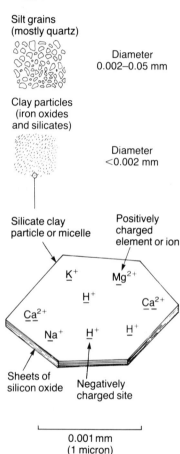

Silicate clay
particle or micelle

Positively
charged
element or ion

Sheets of
silicon oxide

Negatively
charged site

0.001 mm
(1 micron)

Figure 1.5). The smaller silt grains have a diameter of 0.002 mm–0.05 mm while the coarser sand particles have a diameter range of 0.05 mm–2.00 mm. The sand and silt are chemically rather inactive or inert, since quartz grains have little power to hold water or nutrients.

(b)   Clay

Clay exists in soils as very fine particles with diameters of less than 0.002 mm. The clays are of two main types (Figure 1.4). (1) Some clays are amorphous, i.e. they lack a definite structure, and chief of these are the hydrous oxide or *sesquioxide* clays of iron and aluminium. They are common in the soils of the tropics but are also found in other regions. (2) The *silicate* clays are characteristic but not exclusive of temperate soils. They have a distinctive plate-like appearance (see Figure 1.5). Each clay particle or *micelle* is composed of a series of thin sheets of silicon oxide stacked one on top of the other. This arrangement of silicon sheets forms what is called a 'lattice structure'.

The clay particles have a number of important physical and chemical properties. They can absorb two-to-three times their own volume of water. As a result clays swell when wet and shrink when dry. Clay particles also have an ability to attract and hold chemical elements or plant nutrients (see next section). Because of the internal molecular structure of clay particles, they expose a host of negative charges over their outer surface. Such negatively charged sites are able to hold positively charged chemical elements, as shown in Figure 1.5. When chemical or mineral elements are coupled to the clays in this manner, they are described as being *adsorbed* to the clay.

(c)   Chemical salts (*plant nutrients*)

A third product of rock weathering is chemical salts. Such chemical salts are a fairly diverse group (see page 17). They include substances such as calcium sulphate ($CaSO_4$) and magnesium phosphate $Mg_3(PO_4)_2$. Many are soluble and become dissolved in soil water, forming what is called the *soil solution*. When the soluble salts become dissolved, they separate into their respective chemical elements or atoms. For instance, when common salt (sodium chloride: NaCl) dissolves in water it separates into positively charged atoms or ions of sodium (Na^+) and negatively charged atoms or ions of chlorine (Cl^-). The positively charged ions, called *cations*, are attracted to and held by the clay particles, as described in the previous section. The negatively charged ions are known as *anions*. A group of chemical elements, including calcium (Ca), potassium (K) and phosphorus (P), are essential for plant growth and are referred to as *plant nutrients*.

2.   Soil texture

(a)   The balance of mineral particles

Soil texture is determined by the proportion and amount of sand, silt and

Table 1.1 Some soil textures classified according to the percentage amount of sand, silt and clay

Soil textures	Percentage sand	Percentage silt	Percentage clay
Sand	90	5	5
Loamy sand	85	10	5
Sandy loam	65	25	10
Loam	45	40	15
Silt loam	20	60	20
Clay loam	28	37	35
Clay	25	30	45

Figure 1.6 Using a soil textural triangle: (a) the main textural classes; (b) percentage arrangement of sand, silt and clay; (c) defining an example of a sandy loam

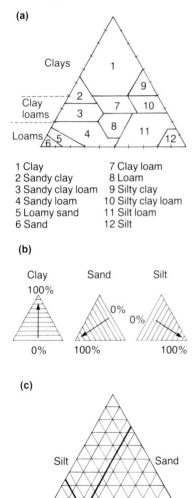

1 Clay
2 Sandy clay
3 Sandy clay loam
4 Sandy loam
5 Loamy sand
6 Sand
7 Clay loam
8 Loam
9 Silty clay
10 Silty clay loam
11 Silt loam
12 Silt

10% clay
25% silt } Sandy loam
65% sand

clay in the soil. Many different soil textures are possible. As shown in Table 1.1, a soil with a high proportion of sand (90% or more) has a sandy texture, while a clay-textured soil has a smaller but still relatively dominant amount of clay (45% or more). A loamy-textured soil or a *loam* has a more even distribution of sand, silt and clay. We can identify and name the texture of a soil if we know the relative amounts of sand, silt and clay that are present. This is done by using a textural triangle of the type shown in Figure 1.6. For instance, if a soil has 10% clay, 25% silt and 65% sand, we classify it as a sandy loam, i.e. a loam with a high proportion of sand.

(b) The importance of texture

Soil texture is very important as it affects many of the factors which influence plant growth and agriculture.

(i) *Moisture content and aeration.* Soil texture affects the water content and drainage ability of soils. This is because texture controls the nature of a soil's *pores*, i.e. the voids or spaces between the mineral particles. In a clay soil, for example, there are very many minute pores or *micropores* between the tiny clay particles. Being small they tend to retain water but to exclude air. As a result, clay soils are prone to drain poorly and to become waterlogged.

By contrast, sandy soils are dry soils. They have fewer but larger pores between the individual grains. These *macropores* hold air but not water, thus allowing the water to drain freely through the soil.

A loam has both sand and clay and so has macropores and micropores, giving a more balanced supply of both air (in the macropores) and water (within the micropores).

Figure 1.7 helps to show how soils retain water and how they drain. After a rainstorm water percolates into a soil, filling or *saturating* all pores with moisture. After a short time, water drains from the larger macropores. Because this water is removed by the force of gravity, it is called *free* or *gravitational water*. At this stage the soil still retains the maximum amount of water possible against the force of gravity. The water is held in the micropores by tension or 'suction' forces. The moisture condition of the soil is now said to be at *field capacity*.

11

Figure 1.7 Moisture conditions in the soil. Moisture levels are indicated by soil saturation, field capacity, and wilting point; types of soil water by gravitational, capillary and hygroscopic water.

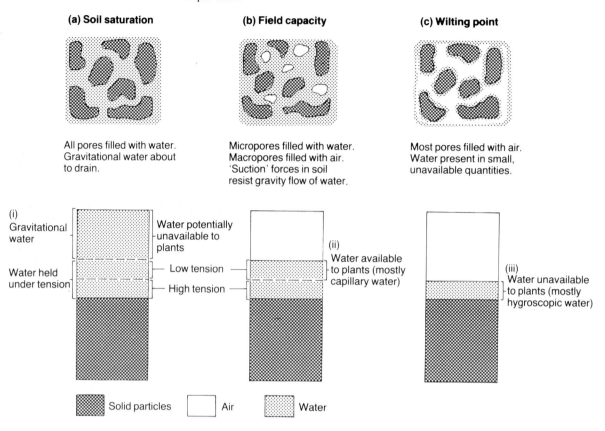

The moisture held at field capacity is of two kinds. A portion of it is held loosely in the soil and can be withdrawn and used by plants. This is the *available* or *capillary water*. The remainder is not available to plants, being held against the surfaces of the soil particles by very strong forces. Most of this unavailable water is called *hygroscopic water*.

After plants have used all the available capillary water from the soil and only unavailable (mostly hygroscopic) water remains, plants continue to lose water by evaporation and *transpiration* (water-flow through the plant). As a result, they lose too much moisture and wilt. At this stage the soil *wilting point* is reached.

(ii) *Nutrient retention.* Soil texture affects the ability of a soil to supply nutrients to plants. For example, coarse-textured soils are lacking or deficient in plant nutrients. The reasons for this are: (1) Sandy parent materials are unable to adsorb cations. (2) Sandy soils allow water to percolate and drain freely, thus helping to wash out salts. (3) Sandy soils often have very little organic matter (a source of nutrients) because the well-aerated condition of the sandy soils helps organic remains to decompose rapidly. Coarse-textured soils are therefore often described as 'hungry' because they need large doses of fertilisers (both organic and

Plate 1.1 Ashdown Forest, east Sussex. This area is heathland on sandy, acid podzolic soils which are lacking in nutrients. There is a mosaic of types of vegetation including moist heathland, dry heathland, purple moor grass and bracken. (Photograph: M. Pulsford, British Geological Survey)

inorganic) to enable crops to be grown successfully. Good examples of sandy, nutrient-deficient soils are the lowland heathland soils of south and east England (see Plate 1.1) and those soils which develop on coastal sand dunes (see Plate 3.6 on page 75).

Clay-rich soils do have a high capacity for adsorbing nutrients but, as already shown, they are prone to waterlogging and oxygen deficiency. Because of this, plants may be unable to use the nutrients that are present in the wet clay.

The most favourable texture for nutrient supply is, once again, the loam. Loamy soils not only have enough clay for holding nutrients but also have enough sand to avoid the worst effects of waterlogging.

(iii) *Ease of cultivation and root penetration.* Soil texture affects how people and plants can use the soil. Light sandy and loamy soils are easily ploughed and cultivated. Heavy clay soils are difficult to work: when dry, they shrink to a hard, impenetrable mass; when wet they become too 'plastic' (like moist putty) for successful cultivation. Plant roots too can penetrate much more easily into light-textured soils than into heavy-textured soils. Many root crops, such as spring potatoes and carrots, are often grown on warm, light, sandy soils such as those found on raised beaches (e.g. along the Ayrshire coast).

ASSIGNMENTS

1. (a) *Refer to Figure 1.5. Compare sand and clay particles with reference to: (i) overall size; (ii) shape and composition; (iii) ability to hold water and nutrients.*
   (b) *Give an example of a chemical salt.*
   (c) *What happens to chemical salts when they dissolve in the soil water?*

2. (a) *Define soil texture.*
    (b) *Give three properties of soils which are influenced by texture.*
    (c) *Compare sandy, clay and loamy soils in relation to these three properties.*

C. Organic Matter

1. Surface accumulation and humus

If the 'physical skeleton' of a soil is the mineral matter, then its 'flesh' is represented by organic matter, which is added to it in the form of dead plant and animal remains and the waste products of living organisms. It is supplied largely to the top layer of the soil, either on the surface itself by leaf- or litter-fall, or within the upper zone of the soil by roots. Millions of soil animals, such as earthworms and insects, as well as fungi and microscopic bacteria live in the soil, breaking down these plant and animal remains. The soil organisms which feed on the dead organic matter can only decompose plant material and mix it with the mineral soil near the surface. As a result, the uppermost horizons of a soil are generally richer in organic matter and, because of this, darker in colour than the lower mineral ones (see Plate 2.1 on page 73). Organic horizons are denoted as A-horizons or as O-horizons if the organic remains are little decomposed (see page 190).

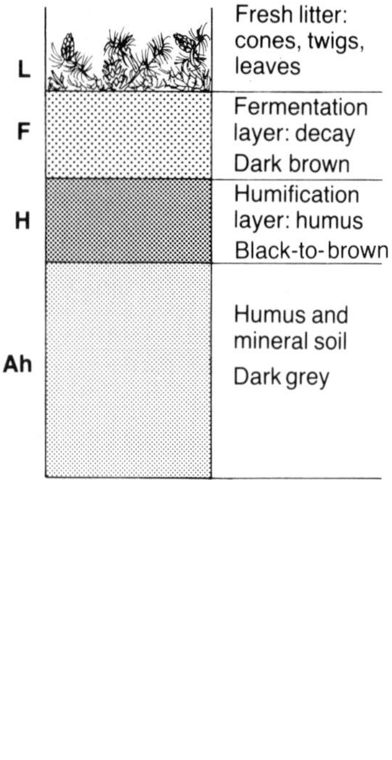

There are often three distinct organic layers at different stages of decomposition at the soil surface (see Figure 1.8). The uppermost or litter layer is made up of freshly deposited, easily recognisable plant remains (leaves, twigs, cones) and is denoted by the symbol L (L = litter layer). Immediately below the litter layer is the fermentation layer (F), where decay is active and where only a few remains can still be identified. The lowest layer is where decay is complete and all the original plant tissues are unrecognisable. This is the zone where *humus* is found. Humus is an end-product of decay and is a black-to-brown, structureless or jelly-like substance. This lowest layer is the H or humification (humus-making) layer.

While litter is being broken down by the soil organisms to form humus, inorganic or mineral materials are released. These include carbon dioxide, some water and some chemical salts. The chemical salts supplement those released by rock weathering but there is one important difference between the two types. Many chemical salts which are formed from decaying organic matter are rich in nitrogen. Along with other nitrogen-rich compounds produced by certain bacteria in the soil, these salts are essential for plant growth.

Figure 1.8 Organic layers at the soil surface showing litter (L), fermentation (F) and humification (H) layers. Where humus becomes mixed with the underlying mineral soil, an Ah-horizon occurs.

L		Fresh litter: cones, twigs, leaves
F		Fermentation layer: decay. Dark brown
H		Humification layer: humus. Black-to-brown
Ah		Humus and mineral soil. Dark grey

2. Types of humus

Different types of humus are also found *between* different soils. They include mull, mor, and moder together with peat.

Mull is a dark-brown-to-black, well-decomposed, crumbly type of humus which is rich in nutrients and only slightly acidic (pH = 5.5–6.5;

14

see page 18). In the A-horizon it becomes thoroughly mixed with the underlying mineral matter by means of earthworms and other soil fauna. Mull forms in well-aerated soils, such as those associated with well-drained lowland areas of the UK which are under deciduous woodland and fertile grassland. The brown soils outlined in Chapter 2, Section B.2 are mostly mull-humus soils.

Mor is a black, poorly decomposed, acid (pH = 3.5–4.5) type of humus which is lacking in nutrients. Unlike mull, mor-humus does not become intimately mixed with the underlying mineral matter, because there are very few earthworms in acid mor soils. Instead the mor forms a cap or mat over the mineral matter below. Being common in wet, acid environments, mor tends to be associated with upland and lowland heathland areas and areas under coniferous forest plantations.

A transitional type of humus with characteristics between mull and mor can also be recognised. *Moder* is the name given to humus which resembles a mor-like mull. It can form, for instance, when deciduous woodland develops on acid, nutrient-deficient parent rocks such as sandstone or granite (pH = 4–5).

In the UK, in flat upland areas and lowland plains, especially where drainage is impeded, soils become more or less permanently waterlogged. Lack of oxygen slows down organic (bacterial) decomposition to a very low level. As a result very thick layers of poorly decayed plant remains gradually build up and a humic deposit known as *peat* develops.

3. Humus: the vital ingredient

Without decaying organic matter the topsoil could not support much plant life. Humus is important to soils and plant growth for the following reasons.

(i) It helps the topsoil to hold the water and oxygen that are necessary for plant growth. Humus can hold much more water than clay, for instance.

(ii) As with clay, the fine organic particles of humus are able to adsorb basic cations on their surface and to release them for plant growth.

(iii) During organic decay and the formation of humus, chemical salts are released which can provide soil nutrients.

(iv) Humus is able to bind together individual mineral particles into larger, combined structures called *aggregates*. These soil aggregates provide a very good material for plant growth (see next section). They are also less easy to remove by wind and rain than are the individual soil particles. Humus therefore helps to stabilise the surface soil layers, so that plants can gain a foothold.

D. Soil Structure

Under natural conditions the individual particles of a soil tend to stick together so that, when a soil is pressed between the fingers, it breaks apart into various-sized clumps. The arrangement of these clumps (also

Figure 1.9   Types of soil structure (Source: Agricultural Advisory Council, 1970)

Size class	Type of structure	Description of aggregates	Appearance of aggregates (peds)	Common horizon location
1 mm–6 mm	Crumb	Small, fairly porous spheres; not joined to other aggregates		**A** - horizon
1 mm–10 mm	Platy	Like plates; often overlapping which hinders water passing through		Plough pan **Ea**-horizon in forest and claypan soils
5 mm–75 mm	Blocky	Like blocks; easily fit closely together; often break into smaller blocks		**B**-horizon: heavy clays, e.g. London Clays
10 mm–over 100 mm	Prismatic	Column-like prisms; easily fit closely together; sometimes break into smaller blocks		**B/C**-horizon: heavy clays, e.g. carboniferous clays
10 mm–over 100 mm	Columnar	Like columns with rounded caps; easily fit closely together		**B/C**-horizon in alkali soils

called aggregates or *peds*) give the soil a particular *structure*. The sizes and shapes of the aggregates vary very much, as shown in Figure 1.9.

The best structure for plant growth is the crumb structure. Small crumbs, about 3–6 mm in diameter, provide an excellent balance between air, water and nutrients, all of which are essential for plant growth. The crumbs are a mixture of fine particles of organic matter and clay and larger grains of silt and sand. The very fine, or *colloidal*, particles of clay and organic matter combine to form what is called the *clay–humus* or *colloidal complex*, which is particularly important because it adsorbs and holds plant nutrients. The crumbs, which are porous, have air-filled macropores and fine, water-retaining micropores.

Farmers seeking to create a soil with a good crumb structure are aware of a number of factors which help this process. These are: changes in heat and moisture along with a system of dense grass roots; calcium which can make clay particles combine, or *flocculate*; but best of all, a good supply of humus in the soil.

ASSIGNMENTS
1. (a)   *What is humus?*
    (b)   *Show how humus can vary: (i) within soils; (ii) between different soils.*
    (c)   *Describe three important effects which humus has on the soil.*

2. (a) *Distinguish between soil texture and soil structure.*
    (b) *Using Figure 1.9, describe five types of soil structure.*
    (c) *Describe how a crumb structure benefits soils and plant growth.*

E. Soil Chemistry

1. Plant nutrients

Of the 92 chemical elements which occur naturally, less than 20 are essential for plant growth and these are classified according to whether plants need large or small amounts of them. The *primary elements* of carbon (C), hydrogen (H), oxygen (O) and nitrogen (N) are needed in large amounts; the *secondary elements*, including calcium (Ca), magnesium (Mg) and potassium (K), are needed in smaller quantities; the *trace elements*, including cobalt (Co) and molybdenum (Mo), are needed in very small quantities.

Many soil nutrients originate as chemical salts which are released by rock-weathering and organic decay. Also, some nutrients are supplied by rainfall, others by the addition of fertilisers. As shown in Figure 1.10, most of these chemical salts dissolve in the soil solution, producing plant nutrients in the form of positively charged ions or basic cations. The basic cations (e.g. Ca^{2+}, Mg^{2+}, K^+) are stored in the soil solution itself or are adsorbed to particles of clay and humus. Cations can be *exchanged* between the particles of clay and humus and the soil solution. Cations may also be removed (exchanged) from the surface of the soil particles or from the soil solution by the action of plant roots. They may be washed from the soil solution in drainage waters. *Cation exchange capacity* (CEC) is a measure of the ability of a soil to keep positively charged ions (cations). Thus, a soil with a low cation exchange capacity has only a limited ability to keep these essential plant nutrients.

Figure 1.10 Origin, distribution and movement of plant nutrients (basic cations) in the soil

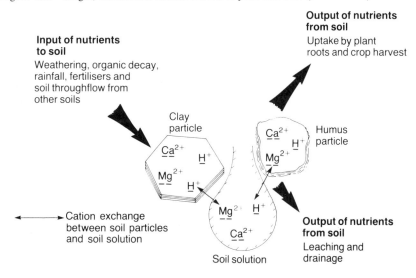

17

Figure 1.11   Acidity level as measured by the pH scale 0–14

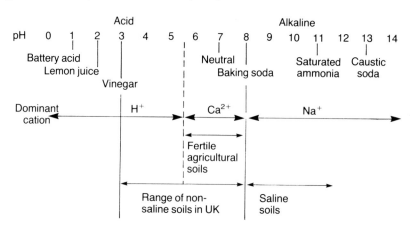

Figure 1.12   Leaching of basic cations such as calcium by acid rain water. It acidity is severe enough, leaching may include the removal from the soil of other solutes, e.g. iron compounds, dissolved clay, soluble organic matter.

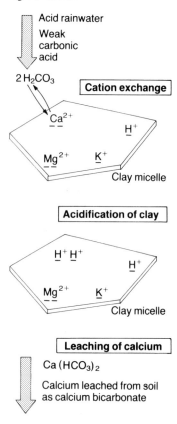

2.   Soil acidity

(a)   Definition

As well as accumulating basic cations such as Ca^{2+}, K^+, Mg^{2+}, the soil also contains positively charged ions of hydrogen (H^+). Hydrogen ions are responsible for the acidity of soil.

Acidity is a measure of the concentration of hydrogen ions in a solution. The degree of acidity (or alkalinity) of a solution is measured on a pH scale, as shown in Figure 1.11. The scale ranges from 0–14: 0 is the most acid (battery acid has a pH of 1); 7 is neutral (the pH of distilled water); 14 is the most alkaline (a saturated solution of ammonia in water has a pH of 11). The scale is logarithmic, which means that each step of a single number increases or decreases acidity by a factor of 10. So a solution with a pH of 6 is 10 times more acidic, i.e. contains 10 times more hydrogen ions, than a solution with a pH of 7. Similarly a solution with a pH of 5 is 100 times more acidic than one with a pH of 7. A solution with a pH of 9 is 10 times more alkaline than one with a pH of 8.

Rainfall is normally on the acid side with a pH of 4.6–5.6. This is because rain absorbs atmospheric gases such as carbon dioxide (CO_2), sulphur dioxide (SO_2) and nitrogen dioxide (NO_2) and forms weak carbonic (H_2CO_3), sulphuric (H_2SO_4) and nitric (HNO_3) acids. Organic acids released during plant-decay also add to the acidity level of rain as it percolates through the soil.

(b)   Leaching

Acidic rainwater moving through the soil is able to remove the basic cations from the soil solution and also those adsorbed to the clay and humus particles. In Figure 1.12, water containing weak carbonic acid (H_2CO_3) drains through the soil and displaces calcium from a clay micelle, replacing it with hydrogen on an exchange site. The calcium is then washed from the soil in solution in the form of calcium bicarbonate in the drainage waters.

The removal of basic cations from the soil by acid rainwater is known as *leaching*. Leaching results in the accumulation of hydrogen ions on the soil exchange sites and so the soil becomes more acid.

(c) Consequences of high acidity

High acidity has a number of effects on the soil: (1) It aids the removal of calcium and other bases from the soil. (2) It makes certain elements such as iron and aluminium excessively soluble. These then become toxic (poisonous) to plants and animals. (3) It makes other elements, such as phosphorus, insoluble. As a result, these become unavailable to plants. (4) It causes the disintegration of clay particles, thereby releasing more iron and aluminium into the soil. (5) It makes soil organic matter become soluble. Like the dissolved clay, this may be washed down the profile and deposited at depth or moved out of the soil altogether.

It seems clear that high acidity should be avoided if soils are to maintain their structure and their ability to support plant growth. A neutral-to-slightly-acid soil, with a pH range of 5.5–6.5, is ideal for crop growth. At this acidity level, there is sufficient calcium to promote clay aggregation and to counteract the souring effect of soil acidity. A pH of 6.5 helps to provide a stable (i.e. non-disintegrating) clay-humus complex and a good soil structure. It also means that nutrients are present in adequate amounts and proportions and are *available* for plant uptake.

F. Soil Fertility: the Constituents Combined

'Soil fertility' expresses how suitable the total soil environment is for crop growth. When soils are fertile, plant growth is rapid and crop yields are high. By contrast, infertile soils are unable to sustain agriculture and so give low crop yields. A fertile soil should supply crops with air, water and nutrients in correct amounts and proportions and should provide a suitable anchorage for root growth. A fertile soil is therefore one which has any or all of the following features:

(i) a sandy loam or loamy soil texture giving good drainage and aeration, as well as an adequate water-holding or field capacity;

(ii) a well-developed crumb structure associated with adequate quantities of soil humus;

(iii) a good supply of nutrients from either organic decay or the addition of fertilisers;

(iv) a neutral-to-slightly-acid pH;

(v) a relatively deep soil providing for root development and ensuring enough reserves of water and nutrients.

Note that achieving high agricultural yields is dependent also on favourable climatic conditions and on farming techniques, including the correct choice of crop to suit the soil and climatic environment.

The idea that soil fertility can be affected by factors external to the soil, in particular by human activity, is now examined.

1. The human impact

Humans can affect soil fertility in two ways. They can increase fertility levels well above those designed by nature. Or they can decrease soil fertility – can even destroy it completely – by misusing and over-exploiting the land.

(a) Increasing fertility

(i) *Traditional or organic farming.* When crops are growing, they remove nutrients and organic remains from the soil. If the crops are harvested and no nutrients are put back into the soil before the next crops are sown, soil fertility and crop yields will decline. Traditional farming methods use two techniques to maintain soil fertility levels. (1) Arable farming is closely linked to livestock farming. Animal manure is mixed with straw to make farm yard manure (FYM) and this is ploughed back into the soil. FYM restores soil fertility by improving soil structure and, as it decays, by returning nutrients to the soil. (2) Crop rotation ensures that nutrient-demanding grain crops (e.g. wheat and barley) are alternated each year with soil-restoring crops, such as grass leys and legumes (e.g. peas). Grass leys add organic matter while legumes restore nitrogen to the soil.

(ii) *Intensive or chemical farming.* Since 1945 traditional agriculture in the UK has become increasingly intensive (see Figure 1.13). The chemical fertility of soil is now maintained and indeed increased by the use of what

Figure 1.13 Intensification of British agriculture shown by the declining numbers of agricultural workers and horses, coupled with the increasing use of tractors and nitrogen-rich, inorganic fertilisers (Source: Newson and Hanwell, 1982, and Agricultural Abstracts, 1983)

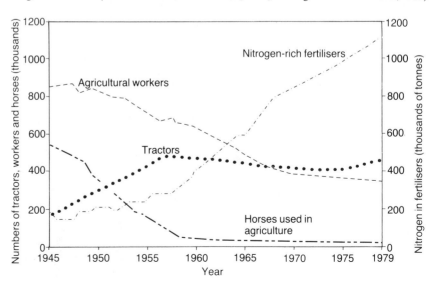

Plate 1.2 Traditional downland showing the scarp (steep) face of the South Downs, west Sussex. There is a rich mix of small fields, old hedges, lanes and copses. (Photograph: Kenneth Scowan)

Plate 1.3 Modern downland field in southern England used for large-scale cereal production. Wild and uncultivated habitats are now confined to narrow marginal strips, chiefly in wet valleys and on steep slopes (Photograph: Farmers' Weekly)

Table 1.2 Trends in agricultural land-use and production in the UK, 1955–1979

		1955	1969	1979
1.	**Cropland** (thousand ha)			
	Arable land	4573	4940	5031
	Temporary grass	2434	2307	1965
	Permanent grass	5477	4997	5104
	Rough grazings	6829	6849	6333
2.	**Numbers of livestock**			
	Total cattle	10 668	12 374	13 426
	Total sheep	22 949	26 604	31 446
3.	**Crop yield**			
	Wheat (tonnes/ha)	3.07	3.92	5.38
	Barley (tonnes/ha)	2.93	3.46	4.24
	Milk (litres/cow)	3187	3741	4653
	Total output			
	Wheat (thousand tonnes)	2725	3689	7311
	Barley (thousand tonnes)	2608	8155	9937
	Milk (million litres)	7569	12 153	15 278

Source: Munton, 1983

are called 'energy subsidies'. These involve applications of large amounts of energy in the form of inorganic fertilisers (rich in nitrogen, phosphorus and potassium), pesticides and herbicides, machinery, water and seeds. Chemical fertilisers are now applied in such large quantities that single crops, such as wheat, barley or oil-seed rape, can be grown in the same field year after year. These single-crop systems are known as *monocultures*.

In addition, by removing hedges, fields have been enlarged to accommodate the use of large modern machinery. This has produced the new 'prairie landscapes' that are now so typical of southern and eastern Britain (*cf.* Plates 1.2 and 1.3). Other signs of land improvement can be seen. Formerly poorly drained soils have been drained and nutrient-deficient heathland has been ploughed and fertilised. Land previously difficult to plough, such as on steep slopes, has been cultivated by using new machinery (e.g. dragline ploughs).

The effects of modern, intensive agriculture on the yields of British crops have been spectacular. Table 1.2 shows that between 1955 and 1979 the yield of wheat has almost doubled while that of barley has increased by nearly 50%. If a fertile soil is defined as one which can sustain satisfactory yields of crops, then modern, high-energy farming has indeed been responsible for increasing the resource, or fertility, value of the soil.

(b) Declining fertility

Under today's intensive agriculture, increases in soil fertility can be achieved only by the constant application of fertilisers and other energy subsidies. Such soil improvements may be illusory, in the sense that many British soils now have little natural fertility. High-energy farming can rob the soil of its natural fertility and can damage soils in the following ways.

(i) *The excessive use of inorganic fertilisers.* Though rich in nutrients, inorganic fertilisers do not contain the organic remains which the former widespread use of FYM once provided. As a result, modern soils are being robbed of a crucial source of humus. This humus is necessary for building up soil structure and for keeping reserves of nutrients and water.

(ii) *The excessive use of pesticides.* Pesticides are environmental poisons and can pose a threat to local soils and wildlife. Many pesticides such as DDT build up in soils and in the tissues of living organisms until they become toxic. As a result, they can kill or do harm to species other than those which they were meant to affect. Many beneficial soil organisms, which are involved in processes of organic decay and nutrient cycling (see page 100), are damaged and killed in this way. Few escape, since 95% of the cropland of the UK is sprayed with some sort of pesticide.

(iii) *Remaining impurities.* Fertilisers and other agricultural chemicals contain impurities. As a result, heavy metals, e.g. lead (Pb), zinc (Zn) and arsenic (As), as well as trace elements, e.g. molybdenum (Mo) and cobalt (Co), may build up to harmful levels in the soil.

(iv) *Cultivation and cropping.* Cultivation can impair soil fertility in three different ways. (1) It causes a loss of soil humus, as organic matter tends to be 'burned up' or oxidised by constant ploughing and aeration. Because of this, soil structures are often weakened. Figure 1.14 shows

Figure 1.14 The stability of soil structure (soil aggregates) under different land-uses (Source: Low, 1972)

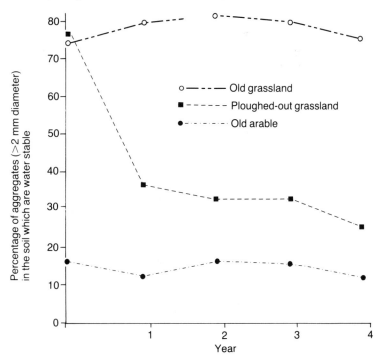

that the longer a soil has been cultivated, the weaker its structure becomes. Structural units (crumbs) break down more easily in long-cultivated soils than in non-cultivated soils under old pasture. When weak structures disintegrate, their individual mineral particles disperse. These dispersed mineral grains clog the soil pores and impede the circulation of air and water through the soil. (2) Nutrients are removed from the soil by growing crops and harvesting them. If nutrients are removed by cropping at a faster rate than they are replaced by fertiliser or FYM, then soil fertility will be reduced. (3) Cultivation leaves the land bare for long periods of time. This exposes the soil to the ravages of wind and rain. Cultivated soils are therefore more prone to erosion and a loss of soil fertility by this means.

(v) *Heavy machinery*. Under pressure from modern heavy machinery, soil structures may become compacted and deformed. This occurs particularly when soils are wet, for it is then that structural or particle bonding is weakest. In Figure 1.15a, a smeared or deformed structural layer is visible immediately below the ploughed zone. This 'plough pan' has flat or plate-like structures which have replaced the free-draining, blocky or crumb structures of the unploughed soil. Soil compression like this reduces the size and amount of pore spaces in the soil. As a result, the bulk density (see Chapter 8, Section B.2) of the soil increases (see Figure 1.15b). Drainage and aeration are impeded which means that waterlogging can occur in the zone above the plough pan.

Figure 1.15   Effect on: (a) soil structure and permeability of ploughing wet soil; (b) soil density of using a tractor (Source: (a) Agricultural Advisory Council, 1970, (b) Newson and Hanwell, 1982)

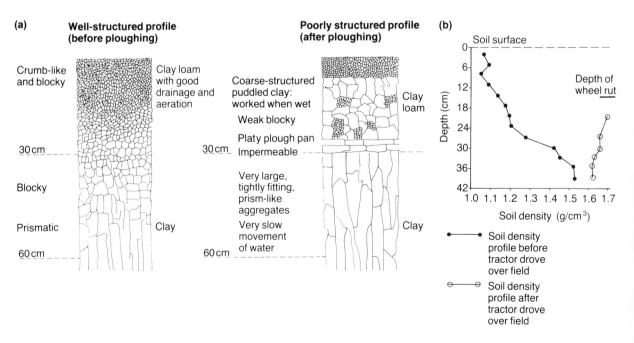

1. (*a*) *Describe what is meant by plant nutrients and show how they are produced in the soil.*

(*b*) *What are the chief characteristics of a fertile soil?*

(*c*) *Illustrate how soil fertility can be increased by: (i) traditional farming techniques; (ii) modern intensive agriculture.*

(*d*) *Describe the way soil fertility can be reduced by: (i) leaching; (ii) increased acidity; (iii) inorganic fertilisers; (iv) over-cropping and cultivation; (v) heavy machinery.*

Key Ideas

A. Introduction

1. Soils consist of weathered surface rocks modified by the addition of dead and decaying plants and animals.
2. Soils are usually described by means of a soil profile which is a two-dimensional, vertical cutting or cross-section.
3. Soil profiles usually consist of layers of differing material called horizons.
4. In addition to having a vertical structure, soils have a horizontal or spatial distribution; soils should therefore be viewed as three-dimensional bodies.
5. Four types of material, often thoroughly intermixed, make up individual soils: mineral particles, organic matter, air and water.

B. The Mineral Particles

1. Three main kinds of mineral matter are found in soils: coarse grains of sand and silt, fine particles of clay, and traces of chemical salts.
2. Two types of clay exist: hydrous oxide clays of iron and aluminium, and silicate clays.
3. The silicate clays, in particular, are able to hold or adsorb chemical elements on their surface.
4. A diverse range of chemical salts, which act as plant nutrients, are made available in soils by rock-weathering.
5. Soil texture refers to the percentage amounts of sand, silt and clay found in soils.
6. By controlling the size and amount of a soil's pores (i.e. the spaces between the mineral particles), soil texture influences the water content and aeration of the soil.
7. Sandy soils have large pores (macropores) and drain freely; clay soils, with many small pores (micropores), tend to suffer from poor drainage and waterlogging; loams have a good balance between small and large pores, giving excellent water-retention and drainage conditions.
8. Only a proportion of the water which percolates into the soil is available to plants; the remainder is lost in drainage or held very tightly against the soil particles.

9. Soil texture also determines the ability of a soil to keep and supply nutrients, and the ease with which roots and the plough can penetrate the soil.

C. Organic Matter

1. Organic matter is supplied largely to the top layers of the soil in the form of dead plant and animal remains and the waste products of living organisms.
2. Soil organisms help to decay dead plant and animal tissues into humus – a black-to-brown, structureless or jelly-like substance.
3. As organic matter decays, chemical salts (plant nutrients) are released.
4. Mull is a well-decomposed, nutrient-rich type of humus; mor-humus is poorly decomposed and acid; peat is a thick deposit of very poorly decayed plant remains.
5. Humus is a potential store of water and nutrients and, by binding a soil's loose particles together into structural units, helps to protect the soil from erosion.

D. Soil Structure

1. Individual soil particles tend to combine together into various-sized clumps, or aggregates, giving the soil a particular structure.
2. Large, blocky and prismatic structures, as well as thin and platey structures, exist but the best structure for plant growth is the crumb.
3. The most effective factor which helps to create a good crumb structure is a good supply of humus in the soil.

E. Soil Chemistry

1. Plant nutrients can be classified into primary, secondary and trace elements, depending on the quantities used by vegetation.
2. Nutrients are present as ions, mostly as positively charged cations in the soil solution and also adsorbed to particles of clay and humus.
3. Cation exchange capacity is a measure of the ability of a soil to hold positively charged ions such as Ca^{2+}, K^+ and P^+.
4. Soil acidity is a measure of the concentration of hydrogen ions (H^+) in solution.
5. Soil acidity is caused by the leaching or loss of positively charged ions (base cations) from the soil and their replacement by hydrogen.
6. High acidity sours the soil, upsets the availability of nutrients and destabilises both clay and humus.

F. Soil Fertility: the Constituents Combined

1. Soil fertility is a measure of the ability of a soil to produce satisfactory yields of crops.
2. Humans can enhance soil fertility by using farm yard manure, crop rotations and inorganic fertilisers.

3. Humans can destroy soil fertility by misusing and over-exploiting the land.
4. Humans reduce soil fertility by extracting nutrients and soil humus faster than they can be replaced, by destroying soil structure and by inducing soil erosion.

Additional Activities

1. Refer to Tables 1.3 and 1.4 and Figure 1.6.
   (a) Describe the percentage amount of sand, silt and clay in the topsoil (A-horizon), subsoil (B-horizon) and weathered parent material (C-horizon) for the three given soil types.
   (b) Using the textural triangle, determine for each soil the soil texture of each horizon.

Table 1.3   Characteristics of three differently textured soils (OD: above mean sea level)

Soil 1 Vale of York: elevation 14 m OD. Land-use: grassland

Soil property	Horizons: A	B	C
Percentage sand	7	10	9
Percentage silt	45	44	55
Percentage clay	48	46	36
Percentage $CaCO_3$	<1	<1	2
pH in water	7.1	7.2	7.7
Percentage organic carbon (humus)	3.6	4.6	1.5

Soil 2 Blakeney Point, Norfolk: elevation 7 m OD. Land-use: lowland heath

Soil property	Horizons: A	B	C
Percentage sand	94	98	97
Percentage silt	2	2	2
Percentage clay	4	0	1
Percentage $CaCO_3$	0	0	0
pH in water	4.8	5.1	5.3
Percentage organic carbon (humus)	2.8	0.7	0.7

Soil 3 Grantham area: elevation 57 m OD. Land-use: arable

Soil property	Horizons: A	B	C
Percentage sand	52	39	84
Percentage silt	29	44	13
Percentage clay	19	17	3
Percentage $CaCO_3$	0	0	0
pH in water	5.7	6.9	7.4
Percentage organic carbon (humus)	0.8	0.3	0

Source: Soil Survey Regional Bulletins, 1984

Table 1.4 Types of soil water in three differently textured soils, expressed as the percentage weight of water to the weight of solid particles

Soil	Field capacity	Capillary water	Hygroscopic water
Sandy soil	7.6	4.2	3.4
Sandy loam	15.5	8.6	6.9
Clay soil	30.4	14.3	16.1

Source: Brady, 1984

Figure 1.16 Area of wheat and barley as a percentage of total area of crops and grass in England in 1955 and 1979 (Source: Munton, 1983)

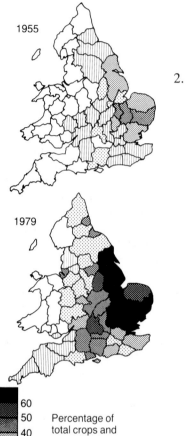

1955

1979

60
50 Percentage of
40 total crops and
    grass that is
30 under wheat and
20 barley combined
10
0

(c) Associate each soil type with one of the following parent materials: coarse, loamy (fluvioglacial) drift; clayey, stoneless, glacial drift; wind-blown sand dunes.

(d) Using Table 1.4 as a guide, examine the relationship between the texture of each soil type contained in Table 1.3 and its probable drainage condition.

(e) Account for the pH level and organic carbon content of each soil horizon.

2. (a) Describe the agricultural trends shown in Figure 1.13.

(b) Describe the changes in the cropland area shown in Figure 1.16 and Table 1.2.

(c) Describe the trends in yield and total production of crops indicated in Table 1.2.

(d) Examine the relationship between the data shown in Figure 1.13 and those in Figure 1.16 and Table 1.2.

(e) Using Figures 1.14 and 1.15, examine how modern farming methods can damage the condition of the soil.

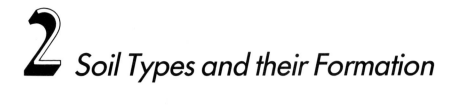

2 Soil Types and their Formation

Introduction: processes and factors

Soil formation may be said to take place when horizons develop within the soil profile. When soil horizons develop they differentiate, i.e. over a period of time they acquire certain characteristics of colour, texture, pH, etc, which distinguish them from neighbouring horizons. One soil type differs from another because each has its own characteristic range of horizons. In order to understand soil formation we need to study the *processes* which allow the differentiation of horizons to take place. These are: weathering, the incorporation of organic matter, and the movements of water.

To know the environment of soils is also important. Each environment (e.g. chalk downland, clay vale, mountain slope) has its own group or combination of *soil-forming factors*, e.g. climate, parent materials, topography (terrain), organisms, and time. These soil-forming factors control the conditions under which the *soil-forming processes* operate, and so they determine the state of any soil, the *soil response*, in any given location.

As shown in Figure 2.1, climate influences the movement of water in the soil and therefore the degree of leaching. For instance, the wetter the climate, the greater the flow of water down through the soil and the more intense the leaching that will result. Because of leaching, some horizons will lose materials (be *eluviated*) while others will gain them (be *illuviated*).

Figure 2.1 Examples of the relationship between soil-forming factors, processes and soil response

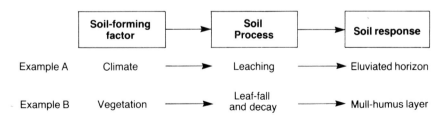

A. Processes of Soil Formation

1. A working model

(a) *Soil as an open system*

Figure 2.2 is a useful model of soil formation which shows the soil as the end-product of a set of processes involving additions, losses, transfers and transformations of materials. The soil is regarded as an 'open' system since materials and energy are gained and lost at its boundaries. Examples of *additions* include water, heat and oxygen from the atmosphere, organic matter from plant and animal remains, and wind-borne dust particles. Materials which are *lost* from the soil include water (by seepage, evaporation, transpiration), mineral particles (by erosion), various chemical elements (by leaching) and organic matter (through decay and decomposition).

Transfers consist of both upward and downward movements. For instance, organic and mineral substances can be moved upwards by soil fauna, e.g. earthworms or moles, or downwards by percolating water. *Transformations* mean changes in both composition and form, and can affect any soil material, whether added, lost or transferred. Organic remains added to the soil are transformed into humus and mineral elements (see page 14) in the process of decay. Transformations of

Figure 2.2 A view of the soil system showing additions, losses, transfers and transformations of materials (Source: Simonson, 1978)

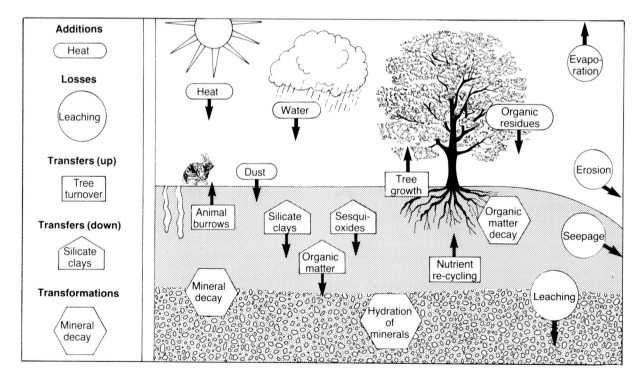

30

silicate minerals contained in rocks like granite and basalt include the breakdown of primary minerals (sand) and the formation of secondary minerals (clays).

(b) *Weathering, incorporation of organic matter and movements of water*

The grouping of processes, shown in Figure 2.2, into gains, losses, transfers and transformations is very generalised. For instance, the loss of organic matter by decay is different from the loss of soluble salts by leaching. We therefore need to reorganise and sub-divide these four processes to have a clearer understanding of soil formation. One arrangement is shown in Table 2.1. The development of soil is seen as the result of three distinct groups of processes: weathering, incorporation of organic matter and movements of water. Each of these can be sub-divided into more specific processes, e.g. weathering into oxidation and *hydrolysis*, water movements into leaching, *gleying* and *salinisation*.

2. Weathering processes

As shown in Table 2.1, the role of weathering in soil formation (*pedogenesis*) is to create an *accumulation of parent materials*, i.e. a collection of fragmented rock and mineral particles. Without even a few centimetres of weathered rock materials, soil formation cannot begin. Weathering controls whether the depth of soil formed will be deep or shallow and influences its colouration, e.g. the red and yellow of oxidised iron compounds or the grey of reduced iron substances within the soil.

Remember, however, that weathering can and does take place *without* the formation of soil. This fact is illustrated by the disintegration of rocks by weathering to depths of at least 25 metres along the Rio Paulo Highway in Brazil.

3. Incorporation of organic material

The first real sign that soils are being formed from weathered bedrock is the incorporation of organic material, which produces distinctive

Table 2.1 Soil processes, divided into general and specific types, and the resulting soil effects

	Primary (or general) process	Secondary (or specific) process	Role in formation of soil
1	Weathering	Oxidation Hydrolysis	Accumulation of parent materials
2	Incorporation of organic matter	Formation of humus Soil aggregation	Main agents responsible for development of horizons within soils
3	Movements of water	Leaching Gleying Salinisation	

surface accumulations of organic remains in the form of A-horizons or A- and O-horizons. The accumulation of soil organic matter varies a lot. Within different soils, there is a gradual change in the composition of organic material as it decays from litter to humus in the LFH-layers of some surface soil horizons (see page 14). There is also a contrast between different soils in terms of the amount, type and distribution of incorporated organic material.

(i) *Amount.* Some soils, especially where decay is rapid (moist tropics) or input is limited (hot deserts, young soils), accumulate very little organic matter (see Figure 2.3). In cool moist climates, where there is waterlogging (e.g. plateaux, gentle slopes in upland Britain), organic decay is slowed down, so that thick layers of peaty surface-humus accumulate.

(ii) *Type.* In parts of lowland Britain which have moderate rainfall, quite high summer temperatures and deciduous woodland or rich grassland, brown soils with mull-humus systems develop (see page 37). By contrast, soils with mor-humus systems are found in north and west Britain in colder, wetter areas under more 'acid' vegetation types, e.g. heather moorland, wet acid grassland, coniferous forest (see page 39).

(iii) *Distribution.* Soils which develop under woodland and forest tend to accumulate organic remains at the soil surface because the principal supply of organic matter is from leaf-fall (Figure 2.3). Grassland soils, on the other hand, have a much more widely distributed organic layer, resulting from both leaf-fall on the surface and the dead remains of a dense and very widespread root system.

Figure 2.3   Differences in the content and distribution of organic matter in the upper 1.2 metres of a grassland soil, a forest soil and a desert shrubland soil. Grassland soils usually contain 50% more organic matter than forest soils, while desert soils have very little organic matter. The structure and distribution of organic matter, and especially that of the overlying vegetation, can be used to identify individual ecosystems. (Source: Tivy and O'Hare, 1981)

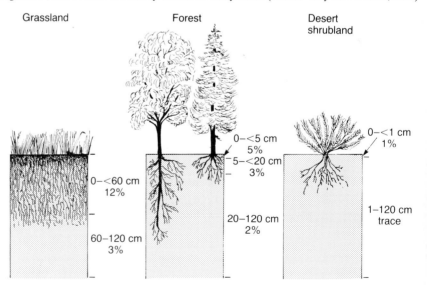

Organic remains can also be redistributed within soils by the movements of fauna. For example, in mull soils the mixture of surface organic and mineral matter is caused by earthworms.

Organic remains and other materials can also be moved and transformed by water in the soil. This is the subject of the next section.

4. Movements of water through the soil

As shown in Figure 2.4, the principal input of water to the soil is from rain (precipitation). When water arrives at the soil surface it can either be lost by immediate surface evaporation or overland flow or else percolate down through the soil under gravity. Of the water which goes into the surface soil layer, a good deal may be drawn back later to the surface by evaporation and, in particular, by plant transpiration. (The total losses by evaporation and transpiration equal *evapotranspiration*). Other pathways by which water is eventually lost from the soil include vertical seepage to groundwater supplies, and lateral downhill seepage in the soil layer. This latter process, known as *throughflow*, may of course add to the soil's water supply, e.g. at the foot of slopes.

The water budget of the soil

The *water budget* of the soil is the term used to express the difference between precipitation input (P) and potential losses from evapotranspiration (Pet), i.e. it is the water potentially available for soil drainage and run off.

Where precipitation input exceeds evapotranspiration losses, there is a positive water budget; where precipitation input is less than evapotran-

Figure 2.4 Movement of water through the landscape and soil

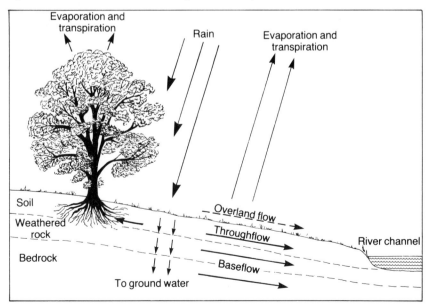

spiration losses, there is a negative water budget. The most important factors affecting the water budget of a soil are climatic conditions and soil drainage.

(i) *Climate and net water flow*. The most significant factor affecting water flow in soils, and thus the movement of materials in soils, is climate.

In areas of high rainfall and low temperatures (e.g. upland Britain), precipitation inputs to the soil (P) will exceed potential losses by evapotranspiration (Pet). Therefore, where the ratio between P and Pet is greater than one (i.e. P/Pet > 1), the resulting or *net* flow of water will be downwards from upper to lower horizons (see Figure 2.5a).

In hot desert areas, which have little rainfall and high rates of potential evapotranspiration loss, there will be a net movement of water upwards from the subsoil to the surface (see Figure 2.5b). But even in the hottest desert regions, where the P/Pet ratio is far less than one (i.e. P/Pet < 1), water cannot easily be transferred from deep *water tables* to the surface by evaporation. However, where the water table is close to the ground surface, this movement may be helped by capillary action (i.e. 'suction'; see Figure 2.5c).

In other regions, e.g. the mid-continental interiors of North America and Eurasia, there is an overall annual balance between rainfall and potential evapotranspiration, i.e. P/Pet = 1. Although water may tend to move downwards in the soil during rainy periods, such movement is reversed during hot, drier periods when the water is drawn back to the surface by strong evaporation. In these areas, there will be little net movement of water either down through the profile or up to the ground surface.

Figure 2.5   The water budget of soils for different climatic regions and drainage conditions. The water budget of a soil is the difference between precipitation (P) and potential evapotranspiration (Pet) and is therefore the water potentially available for drainage and run off.

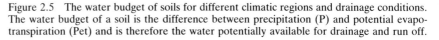

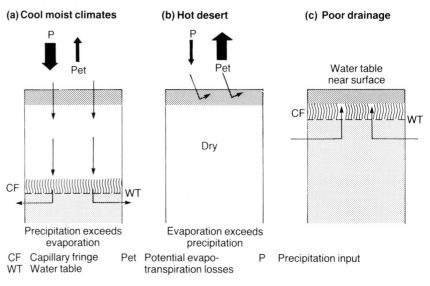

(a) Cool moist climates

P

Pet

CF ||||||||||||||||||||||||||||| WT

Precipitation exceeds evaporation

(b) Hot desert

P

Pet

Dry

Evaporation exceeds precipitation

(c) Poor drainage

Water table near surface

CF)))))) |||||||)))))) |||||))))))) WT

CF   Capillary fringe
WT   Water table

Pet   Potential evapotranspiration losses

P   Precipitation input

(ii) *The movement of soil materials*. The flow of water through the soil affects the transfer of a range of different soil materials, including soluble salts (bases), clay, iron and aluminium sesquioxides, silica and organic remains.

In temperate zones where P > Pet, there is a net downward movement of moisture in the soil. Downward percolating rainwater contains dissolved oxygen and carbon dioxide (from the atmosphere) as well as organic acids derived from decaying plant remains. As outlined in Chapter 1, section E, acid rainwater dissolves and removes soluble salts (bases) from the soil, making the soil even more acid. The term *leaching* (see Figure 2.6a) describes the movement of such solutes in percolating waters.

When the P/Pet ratio increases, especially in the case of acid, very well-drained parent materials (e.g. sand), leaching becomes intense. Many substances in the soil become soluble and therefore subject to removal. These include clay, organic matter, silica, and iron (Fe) and aluminium (Al) sesquioxides. *Podzolisation* is the name for the downward movement, under intense leaching, of most solutes in the soil, but particularly those of iron and aluminium sesquioxides (see Figure 2.6b).

In addition to the chemical movement of substances through the soil, particles can be transferred physically in suspension. When water percolates downwards in the profile, dispersed clay particles may be carried physically in suspension from the upper-surface A-horizon and deposited in the subsoil B-horizon. This process is referred to as *mechanical downwash* (see Figure 2.6c).

A useful general term for the transfer, in any form or direction, of materials in soils is *translocation*. More specifically, the washing out or removal of any soil substance is called *eluviation*; the washing in and deposition of material within soils is called *illuviation*. Soil layers depleted by eluviation are known as eluviated horizons. Illuviated horizons are enriched by processes of illuviation.

(iii) *Restricted water flow*. When the flow of water through the soil is hindered by poor drainage, soils become wet and waterlogged. Soils which become waterlogged restrict the penetration of air or oxygen within the profile. As shown in Figure 2.6d, waterlogging or lack of oxygen has three effects on the soil. (1) It leads to the process of *gleying*, which is the reduction of red-coloured, ferric iron compounds to colourless, or grey, ferrous iron complexes. The most distinctive feature of gleying, however, is the scattering of red patches, or mottles, within the mainly grey gleyed soil. These mottles occur where local pockets of air re-oxidise the ferrous compounds so that they become mottled in the red-yellowish colours of ferric iron. Places where mottles happen are, for example, the larger soil pores, structural cracks or live-root channels, where oxygen has been able to penetrate when the waterlogging is reduced. (2) Gleying encourages the removal and redeposition of the more soluble ferrous iron compounds within the soil. Iron-rich concentrations are often redeposited as layers in the form of *iron pans*. (3) Waterlogging causes a build-up of organic remains at the surface of the soil as a result of decreased organic decay.

Figure 2.6 Soil processes of leaching, podzolisation, mechanical downwash and gleying, in relation to the water budget of soil (Source: Burnham, 1980)

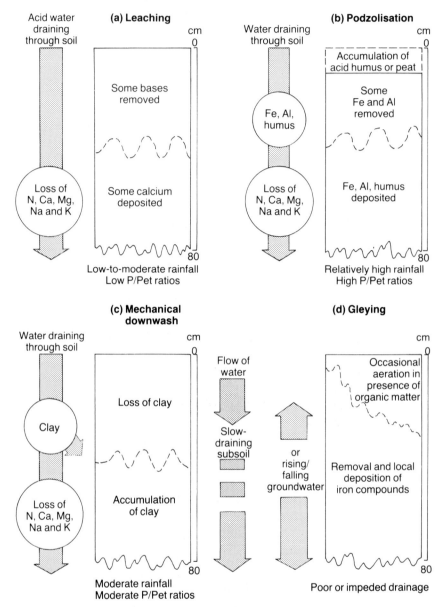

ASSIGNMENTS

1. (a) *Using Figure 2.2, make four separate lists showing additions, losses, transfers and transformations of materials.*

   (b) *Can you add to the lists any other items which are not shown in Figure 2.2?*

   (c) *Name and describe any processes which are not easily classified into one of the lists you have prepared.*

   (d) *Give reasons for your answers.*

2. *Refer to Figures 2.4, 2.5 and 2.6.*
   - (*a*) *What is meant by the water budget of a soil?*
   - (*b*) *Describe at least two main factors which control the water budget of a soil.*
   - (*c*) *Define the following soil processes: leaching, podzolisation, mechanical downwash, gleying.*
   - (*d*) *How are each of these processes related to the water budget of a soil?*
3. (*a*) *Define soil formation.*
   - (*b*) *Examine briefly the roles of weathering, incorporation of organic matter and movements of water in soil development.*
   - (*c*) *Distinguish the difference between soil-forming processes and soil-forming factors.*

B. British Soils

1. Introduction

A great variety of climate, parent materials, topography and vegetation is found in Britain. Similarly the human impact on the landscape is very varied throughout the country. As a result, many soil-forming environments occur: from the wet, lowland bogs of Caithness to the dry, chalk downlands of southern England or the agricultural plains of eastern Britain. This variety of environment has encouraged many different soil types to form. The separate Soil Surveys of England and Wales and of Scotland have identified and mapped no less than 850 different soil types.

Despite this, it is possible to classify the soils of Britain into five main groups: brown soils, podzolic soils, gleys, peats, and various types of 'raw' soil.

2. Brown soils

Brown soils, or brown earths, are described as well-drained, or moderately well-drained, with a reddish-brown horizon which extends below 30 cm in depth. They are a widespread group covering about 45% of the land surface of England and Wales. Four main sub-types are recognised: typical, leached, acid, and calcareous.

(*a*) *Sub-division of brown soils*

(i) *Typical brown soils.* The chief features of the *typical brown soils* are shown in Figure 2.7a and in Plate 2.1 on page 73. The surface A-horizon is biologically active and is denoted by a rich intermixture of mull-humus and mineral matter. The close mixing of the organic and inorganic materials is caused by very active soil organisms, especially earthworms. The mineral B-horizon, which is moderately weathered, is seldom clearly

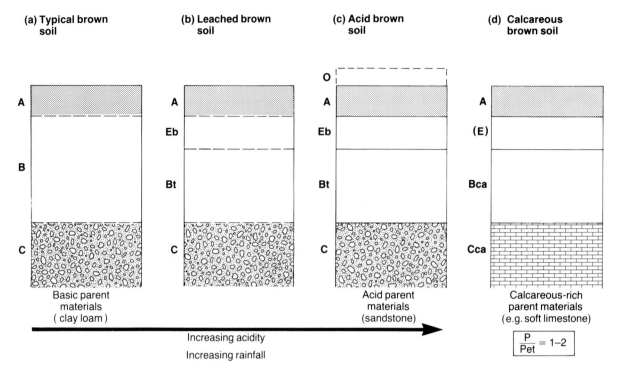

Figure 2.7 Four soils belonging to the brown soils group: (a) typical brown soil; (b) leached brown soil; (c) acid brown soil; (d) calcareous brown soil; and their conditions of formation

separated from the A-horizon, again because of the mixing between the two layers that is done by soil fauna.

The gradual transition between A/B/C-horizons of this soil is also because of limited leaching under low-to-moderate rainfall conditions, i.e. P/Pet ratios of between one and two. Leaching removes only the more soluble bases down through the profile. As a result, these soils are rarely very acid, with characteristic pH levels of 5.0–6.5. Moreover, little or no clay, iron or organic matter, which would otherwise help to distinguish horizons, is washed downwards from the A- to the B-horizon.

(ii) *Leached brown soils*. With slight increases of rainfall and leaching, typical brown soils become *leached brown soils* (Figure 2.7b), in which some translocation, or mechanical downwash, of clay occurs from surface to deeper layers. A clay-rich B-horizon (Bt) is therefore recognisable. Some iron may also be washed down and deposited in this layer. With the removal of some clay and iron from the A-horizon, a weak zone of eluviation (Eb) develops between the A- and the Bt-horizons.

(iii) *Acid brown soils*. On acid parent rocks (e.g. sandstones, granite), moderate rainfall may cause leaching of bases such as calcium and so drop the pH to 4.5–5.5, giving rise to *acid brown soils* (Figure 2.7c). A thin mat of litter (O) is usually found at the surface. This is because there is less intense mixing of organic and mineral matter owing to the lack of earthworms in the more acid conditions. This fact, together with more active eluviation and illuviation, accentuates the separation of the A/B/C-horizons.

(iv) *Calcareous brown soils*. These are defined as non-alluvial, loamy or clayey soils with a weathered calcareous (chalky) subsoil (Figure 2.7d) formed from parent rocks such as limestone. The soils are moderately eluviated (written (E) within brackets) and illuviated. Clay-rich B-horizons may develop, especially on impure, clay-rich limestones. Particles of calcium in the B-horizon (Bca) can increase pH levels to 6.5–8.0.

(b) Distribution and environment

Brown soils are widespread on permeable, non-acid parent materials such as loams and sandy loams. They occur mainly below heights of 300 metres, and so dominate the warmer lowlands of Britain. Brown soils are associated with areas of low-to-moderate rainfall. A distinction can be made, however, between the distribution of leached brown soils, found in the drier lowlands of eastern Britain, and acid brown soils which occur in wetter, lowland environments in the west (see Figure 2.21). Brown soils are essentially mull soils and are related to the more nutrient-rich types of vegetation. They are traditionally associated with broad-leaf, deciduous forest and many now lie below fertile improved grasslands.

(c) Human use

The loamy or sandy texture of brown soils, together with their high content of bases and mull-humus, make them excellent agricultural soils. Good yields are often obtained from them with the minimum of artifical treatment such as fertilisers or drainage.

3. Podzolic soils

Podzolic soils, which have a black, dark-brown-to-grey horizon just below the surface, are characterised by the accumulation of iron and aluminium or organic matter or some combination of both. They are characteristic of the colder and wetter parts of Britain. Like the brown soils they can be sub-divided.

(a) Sub-division of the podzolic group

(i) *The typical podzol*. As indicated in Figure 2.8b, the surface layers of the *typical podzol* are peaty, as a result of poor organic decomposition under cold, wet, acid conditions. An O-horizon with LFH-layers grades into an A-horizon of raw, peaty mor-humus. Below this layer there is an abrupt change to the most distinctive feature of the podzol – a strongly eluviated Ea-horizon composed mostly of bleached, mineral quartz grains (see Plate 2.2 on page 73). The term 'podzol' is a Russian word meaning 'ash coloured', which relates to the colour of the Ea horizon

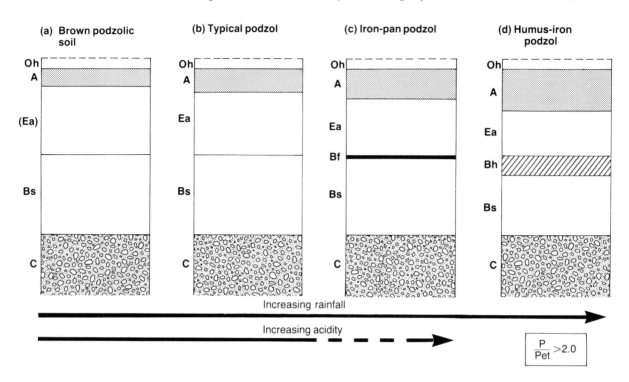

Figure 2.8 Four soils of the podzolic soil group and their conditions of formation

(a) Brown podzolic soil

(b) Typical podzol

(c) Iron-pan podzol

(d) Humus-iron podzol

Increasing rainfall

Increasing acidity

$$\frac{P}{Pet} > 2.0$$

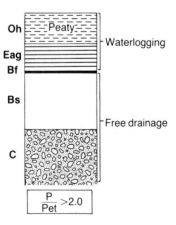

Figure 2.9 Peaty iron pan podzol with gleying

$$\frac{P}{Pet} > 2.0$$

(a = ash). Below the eluviated zone there is a sharp change to the highly illuviated B-horizon, where percolating waters have redeposited various materials from the Ea layer, e.g. iron (Bs), humus (Bh) and clay (Bt).

Having lost most of their bases through strong leaching, podzols are quite acid in reaction, with pH values of 3.5–4.2 being common. Horizon boundaries are sharply defined because of intense eluviation and illuviation processes, and because of limited activity by fauna. The high acidity levels of most podzols greatly restrict the growth of large populations of earthworms, and so soil mixing and the blurring of horizons are slight.

(ii) *Other types of podzolic soil. Humic* or *peaty podzols* have an even thicker layer of raw humus at the surface (see Figure 2.9). They develop in wetter areas with high rainfall. Where humus and iron have been washed down the profile, *humus-iron podzols* occur. As shown in Figure 2.8d, and Plate 2.2, they have dark-coloured humus (Bh) and reddish, iron-enriched (Bs) subsoils.

Iron pan podzols (Figures 2.8c and 2.9) can be identified by the presence of an iron pan (Bf), less than 1 cm thick, usually at or near the top of the B-horizon. Iron pans are composed of mineral grains cemented by high concentrations of iron and humus. When widespread they are often impervious to water, which causes superficial waterlogging of the A-horizon.

Brown podzolic soils are intermediate between fully developed brown soils and podzols. They are generally more acid than acid brown soils, with pH values of 4.0–4.5. As shown in Figure 2.8a, mor- or moder-humus surface layers (O and A) are found with partial, light-coloured, eluvial zones (written (Ea) in brackets) and dark-coloured, iron-rich, illuvial B-horizons.

(b) Stages in podzol formation

A number of stages in podzol formation, or podzolisation, can be identified:

(i) High rainfall acting on permeable, acid parent rocks (e.g. quartzite, sandstone) encourages intense leaching.

(ii) With intense leaching most of the soluble salts, particularly calcium, are removed from the profile, resulting in further increases in acidity.

(iii) High acidity and increased wetness of the soil promote the accumulation of a thick, acid mor (or peaty) layer at the surface.

(iv) During the decomposition of this peaty surface mat, organic acids (e.g. fulvic acid) are released into the soil.

(v) These organic acids, known as *chelating agents*, attack the silicate clays by releasing both iron and aluminium into the soil.

(vi) The chelating agents then combine with the metallic ions (Fe, Al) to form organic-metal compounds known as *chelates*. These are soluble and readily pass down through the profile (by *cheluviation*) and are deposited lower down the soil profile. Iron and aluminium may be deposited in the subsoil because they become less soluble in the slightly higher pH levels found there.

(vii) Organic matter (mor-humus) and some decomposed clay also become soluble under high acidity and are likely to be washed down to lower horizons.

(c) Distribution and environment

Peaty podzolic soils are widespread on reasonably well-drained sites in the colder, wetter uplands of Britain, from Dartmoor through Wales to north-west Scotland (see Figure 2.21). Podzols are associated with acid weathering, however, and typical podzols are found on any acid, excessively drained, sandy parent materials such as loose sands/sandstone or weathered granite. The excessive drainage of these materials helps leaching and podzolisation. Thus, well-developed typical podzols are found on sandy deposits in lowland England, where the climate is both warm and fairly dry, e.g. the Weald sandstone and the fluvioglacial sands of the Breckland, Norfolk. Typical podzols also occur, as shown in Figure 2.21, on granitic parent rocks in the cool but dry climates of north-east Scotland.

Podzols are closely linked to vegetation. They are strongly developed when they lie below 'acid-forming' types of vegetation. Typical podzols are associated, as shown in Plate 1.1 on page 13, with well-drained lowland heaths (e.g. the Weald sandstone, the sands of the Breckland, Norfolk, and the Bagshot sands of Hampshire) and dry, upland heather moorlands. Wet heather moorland, with grasses such as *Nardus* (moor mat grass) and *Molinia* (purple moor grass), gives rise to peaty podzolic soils. Wherever coniferous forests or plantations occur, well-developed podzols are usually found underneath.

(d) Human use

Podzols are not good agricultural soils because they are acid, lacking in bases and generally infertile. The brown podzolic soils can be cultivated, but only with major and sustained effort. This involves, for example, ploughing (to mix the depleted surface layer with enriched deeper layers), frequent liming (the addition of calcium to neutralise acidity) and fertilisation (to maintain a reasonably high supply of nutrients). Untreated podzols are mostly used for rough grazing and coniferous forestry plantations. Since podzols lie beneath much of the wild and semi-wild landscapes of Britain, e.g. heather moorland, mountain grasslands and forest, they may be considered a useful base for recreation.

4. Gleys

Gleys are poorly drained soils which suffer from periodic or permanent waterlogging. They are subject to the process of gleying (see page 35) and so have characteristic grey or red-mottled sub-surface horizons.

(a) Types of gley

When waterlogging is caused by slow-draining subsoils (e.g. clays) *surface-water gleys* occur. These are prominently mottled above a depth of 40 cm. *Groundwater gleys* are found where permeable soils or parent materials (e.g. sands; alluvium, i.e. river-deposited material) are affected by periodic fluctuations in the water table. Seasonal swings in moisture between wet winters and dry summers may induce these fluctuations in the level of the groundwater. Groundwater gleys are often strongly mottled or have uniformly grey subsoils. Despite the dual origin of gleys and the gleying process, almost any soil which does not drain well can be gleyed. For instance, wet, upland, peaty podzols can be gleyed, but so too can many lowland brown soils (see Figure 2.21, and Plate 2.3 on page 73).

(b) Peaty iron pan podzol with gleying

Figure 2.9 shows a profile of a peaty iron pan podzol with gleying. This soil is found in very wet, periodically waterlogged areas of upland Britain (e.g. Dartmoor, Lake District, western uplands of Scotland) and shows most of the symptoms of waterlogging. Because of severe waterlogging under cold conditions, there is a very thick peaty surface layer. The continuous iron pan, which is present at the top of the B-horizon (Bf), is black or dark brown in colour. Being impervious, it affects the distribution of the gleyed horizons (g): above the impervious iron pan, the Ea-horizon is gleyed; below the iron pan, in the remainder of the B-layer, the soil is well aerated, ungleyed and of a bright reddish-brown.

5. Peat (organic) soils

Under highly waterlogged conditions, where peaty surface layers of a soil continue to build up, *peat soils* are found. These are so called if their organic O-horizon is at least 40 cm deep (see Plate 2.4 on page 73). Peats can be categorised on the basis of how they originated and/or their acidity level.

(a) Valley and basin peats

As illustrated in Figure 2.10, valley peats form in poorly drained depressions (lowland or upland), where water running off surrounding higher land accumulates. When nutrients are provided in the drainage from neighbouring calcareous soils, *base-rich* or *fen peat* occurs (e.g. Somerset levels, the Fens). When this type of peat is drained artificially, it becomes a rich agricultural soil, as shown by the widespread cultivation of fen peats for wheat, barley and a variety of horticultural crops (e.g. lettuce, cauliflower, carrots). Many basin peats are acidic, however, since they are fed by streams passing over acid parent rocks (e.g. granites, sandstone). The great basin bog at Rannoch, central Scotland, is a good example of an acid basin peat.

(b) Rain-fed peats

(i) *Raised bogs.* Rain-fed peats depend on high rainfall (high P/Pet ratios) and impeded drainage. As these peats derive most of their nutrients from rainfall, they are normally very acid (pH 3.8–4.4). Over a long period

Figure 2.10 Peat environments showing blanket, valley and raised bogs (Source: Etherington, 1982)

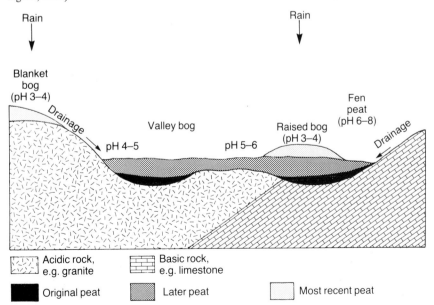

Plate 2.5 Raised bogs at Tregaron, Cardiganshire, Wales. The three bogs shown are composed of very thick layers of acid peat. (Photograph: Cambridge University Collection)

the acid, rain-fed peats, as they continue to accumulate, can develop as *raised bogs* from valley or basin peats. These bogs rise and isolate themselves from the underlying drainage system and depend for their continued existence on high rainfalls. The raised bogs at Tregaron, Wales (see Plate 2.5) well illustrate this type of peat development. As shown in Figure 2.11, a moraine (i.e. a ridge of debris left by a glacier) across the valley of the River Teifi above Tregaron has impeded river flow and so encouraged the saturation of the land and the accumulation of peat.

(ii) *Blanket peat*. The most widespread of the acid, rain-fed peats in Britain is *blanket bog* (see Figure 2.10) which, like the raised peats, bears characteristic plant species including bog or sphagnum moss and cotton grass (*Eriophorum*).

Some people suspect that the extensive blanket peats and bogs and peaty iron pan podzols, which are so characteristic of many flat or gently sloping upland areas of Britain, are associated with human acitivity. This is because these soils lie above older, 'drier' soils, e.g. podzols and brown podzolic soils. Many bogs have been caused by the soil becoming increasingly wet as a result of people reducing evapotranspiration losses from the soil by removing the original upland tree cover. As these areas were already rather moist because of high rainfalls, the now-treeless land became saturated and bog vegetation developed.

Figure 2.11   Raised bogs at Tregaron, Cardiganshire, Wales

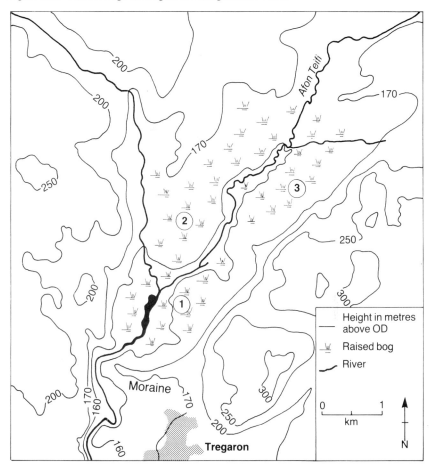

6. 'Raw' soils

These soils (also called lithosols) are found where soil formation has not proceeded very far and so they are shallow soils with a limited development of horizons. Typically they have a thin, organic A-horizon resting directly on the coarser, weathered fragments of bedrock.

Selected types

One group of raw soils occurs in sand dunes (see page 71), salt marshes, etc, where the deposition of parent material has been very recent.

A second category includes *Alpine soils*, which dominate the high mountain areas of Britain above about 1800 metres (see Figure 2.21). These have thin, acid, peaty A-horizons lying above hard bedrock or various screes and mountain 'wastes'. Very cold temperatures in the high mountains, with annual average temperatures of less than 4 °C, prevent deep weathering and the formation of a thick, or biologically active, A-horizon.

Figure 2.12   Rendzina soil

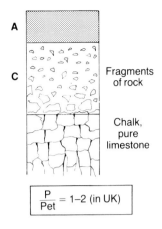

Fragments of rock

Chalk, pure limestone

$\dfrac{P}{Pet}$ = 1–2 (in UK)

The *rendzina* is a raw soil which is found on soft limestone and chalk in association with grassland rich in herbs. (A *herb* is a small, broad-leaved, non-woody plant). In contrast to the calcareous brown soil, it contains few remains of the weathered parent rocks. However, high pH levels resulting from the parent rocks allow thin, dark, but biologically active, A-horizons to form. As shown in Figure 2.12, and Plate 2.6 on page 74, the surface mull-horizon rests directly on the chalky C-horizon.

ASSIGNMENTS

1. *Refer to Figures 2.6–2.9 and Plates 2.1 and 2.2.*
    Compare brown soils with podzolic soils under the following headings:
    (i) climatic conditions and water budgets; (ii) related vegetation; (iii) processes of formation; (iv) profile development; (v) distribution in Britain; (vi) human use.
2. *(a)   Describe the conditions under which soils become waterlogged.*
    (b)   What are the chief features of a waterlogged soil?
    (c)   Define the process of gleying.
    (d)   Refer to Plate 2.3. How can gleying be identified within soils?
    (e)   Describe the soil shown in Figure 2.9. Compare it with those in Figure 2.8.
3. *(a)   Refer to Plate 2.4. What is a peat soil?*
    (b)   Using Figure 2.10, describe four types of peat bog.
    (c)   Refer to Figure 2.11 and Plate 2.5. What type of peat is found north of Tregaron, Wales?
    (d)   Trace the development and changing character of this peat soil over a long period of time.
4. *Study the two soil profile examples shown in Figure 2.13.*
    (a)   Describe the environmental conditions associated with each soil type: relief, slope, altitude, parent rocks (texture), vegetation/land use and possible climate.
    (b)   Describe the sequence of horizons of each soil using: (i) information from the text; (ii) the data given on pH, percentage organic matter, and texture.
    (c)   From the environmental and profile evidence that you have given so far, identify the main processes (e.g. leaching, gleying, humus formation) involved in the formation of the two soils.
    (d)   Name each soil type from the following list: peat, acid brown soil, typical brown soil, peaty iron pan podzol with gleying, humus-iron podzol, rendzina.

C.   Soils and Environment

1.   The soil-forming factors

The principal soil-forming factors mentioned in the introduction to this chapter, i.e. climate, parent materials, topography, organisms and time, are illustrated in Figure 2.14. This arrangement of the factors is rather more complicated than that suggested in Figure 2.1. All five factors are

46

Figure 2.13 Profile analysis and environmental characteristics of two soil types in the UK

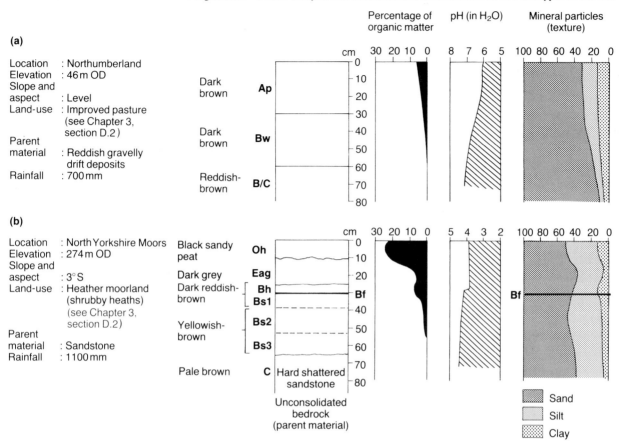

(a)

Location : Northumberland
Elevation : 46 m OD
Slope and
aspect : Level
Land-use : Improved pasture
(see Chapter 3,
section D.2)
Parent
material : Reddish gravelly
drift deposits
Rainfall : 700 mm

(b)

Location : North Yorkshire Moors
Elevation : 274 m OD
Slope and
aspect : 3°S
Land-use : Heather moorland
(shrubby heaths)
(see Chapter 3,
section D.2)
Parent
material : Sandstone
Rainfall : 1100 mm

shown, together with: (i) their effect on the soil; (ii) their effect on each other.

For instance, climate has a *direct* effect on soil processes (e.g. leaching, podzolisation) and hence on soil development over a period of time. However, other factors affect, and are affected by, climate. Climate

Figure 2.14 Main factors in soil formation (Source: Smith, 1984)

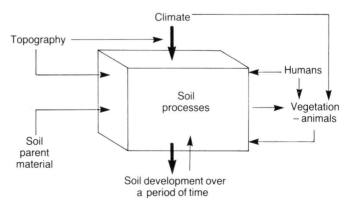

is modified by the topographic factor, which includes the factors of altitude and slope: mountain climates (cold, windy and wet) are very different from valley climates (sheltered, and generally warmer and drier). At the same time, climate influences the distribution of organisms in an area, including the vegetation and the populations of animals and humans. We might say, therefore, that climate exerts an *indirect* effect on the soil via the *biotic factor*.

This illustration of factor action and *interaction* can of course be extended to the other factors of topography, parent materials and organisms, with the exception perhaps of time. They each have a direct and indirect influence on soil formation. The effects of parent materials, topography and time on creating local and regional soil distributions are examined in the remainder of this section. Vegetation is considered in Chapter 3, while the global relationship between soils, vegetation and climate is the subject of Chapters 5 and 6.

2.   Relationships with parent materials

Various types of rock, including solid bedrock (e.g. shales, sandstone, granite) and superficial deposits transported by wind, water and ice (e.g. sand dunes, river alluvium and glacial boulder clay respectively), have an important role to play in soil formation. This role depends firstly on the ability of the original parent rocks to be weathered, secondly on the nutrient content and thirdly the permeability of the weathered remains.

(a)   Rate of weathering

Rocks vary in their resistance to weathering, resulting in different rates and depths of soil formation. An illustration of this effect can be seen in Figure 2.15, which shows a cross-profile of the soils in the Chruch Stretton district of Shropshire. Shale and boulder clay in the valley are soft and easily weathered, yielding deep soils (brown soils and gleys); the relatively hard sandstone and *conglomerates*, as well as the hard *igneous* rock (andesite), give shallow, often stony soils (brown soils and podzols).

(b)   Nutrient content

Different rocks produce weathered remains of very different chemical composition. In Figure 2.15, only the boulder clay contains fragments of calcium carbonate, which gives high pH levels (6.5–7.0) in the B-horizon of soils formed above it. The andesite is fairly rich in bases but, unless it has lime added to it, the pH of its surface layers is quite low (4.5–5.0), giving rise to acid brown earths. Andesite weathers so slowly that any bases which are released by weathering are easily leached away in drainage waters. The sandstones and conglomerates, together with the siltstones, are rich in silica and low in bases. They give strongly acid soils, e.g. podzols and acid brown soils respectively.

Figure 2.15   Effect of parent material and other factors on soil development in the Church Stretton district, Shropshire (Source: Burnham, 1980)

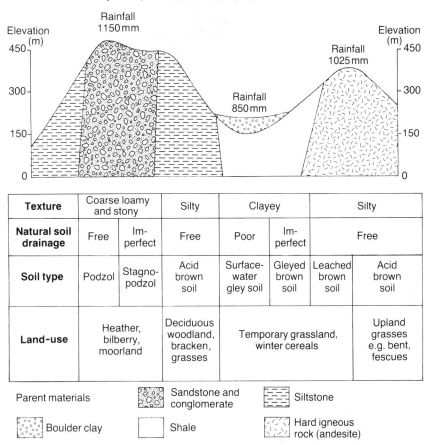

Texture	Coarse loamy and stony		Silty	Clayey			Silty
Natural soil drainage	Free	Im-perfect	Free	Poor	Im-perfect	Free	
Soil type	Podzol	Stagno-podzol	Acid brown soil	Surface-water gley soil	Gleyed brown soil	Leached brown soil	Acid brown soil
Land-use	Heather, bilberry, moorland		Deciduous woodland, bracken, grasses	Temporary grassland, winter cereals			Upland grasses e.g. bent, fescues

Parent materials — Sandstone and conglomerate — Siltstone

Boulder clay — Shale — Hard igneous rock (andesite)

(c)   Texture and permeability

The amount of water in a soil and the speed with which it flows through the soil are influenced by the texture and permeability of the weathered remains. In Figure 2.15, notice that fairly high rainfall (1150 mm), acting on very permeable, coarse-grained sandstones and conglomerates, leads to intense leaching and the formation of a podzolic soil. However, brown soils are found on medium-grained siltstones and weathered andesite. These soils are a result of not only lower rainfall (1025 mm) but also slower drainage and leaching within parent materials of medium permeability.

The almost impermeable, valley boulder clay, which yields clayey soils, is associated with less rainfall (860 mm) but, being low-lying, it is subject to poor drainage and waterlogging. Thus, highly gleyed soils (surface-water gleys) are found where drainage is poor, and less-marked gleys (gleyed brown soils) where drainage is slightly better or imperfect.

3. Soils and topography

(a) The soil catena

Soils and slopes are closely related. Soils vary according to the gradient and position they occupy on a hill-slope. In order to describe this lateral variation of soils on hill-slopes of similar lithology, the term *soil catena* is used.

(b) Why should soils vary along hill-slopes?

Figure 2.16 shows a model of a soil catena which illustrates that there are three reasons for soil variation across the length of a hill-slope.

Figure 2.16   Soil catena showing variation in soil processes and effects along a slope

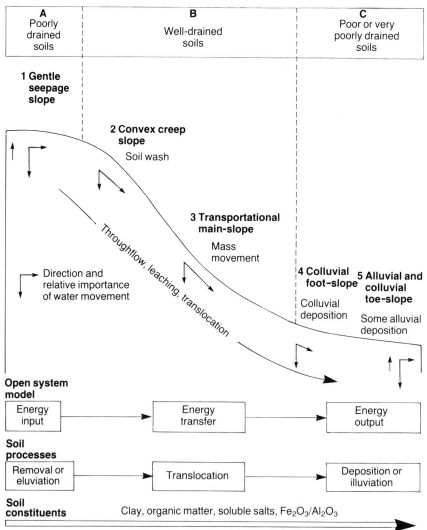

(i) *Drainage condition.* Where drainage is poor, i.e. on upper flat-slopes and lower foot- and toe-slopes, waterlogging tends to occur, resulting in gleyed soils and the development of peat.

(ii) *Transport of eroded material.* Because of the surface movement of sediment down the slope by water action (soil wash) and by the force of gravity (mass movement), deep soils known as *colluvial* soils are found on the lower slopes. Thick soils do not accumulate at hill crests. Only thin, immature, 'raw' soils develop there because soil erosion and soil transport are more effective than weathering, and so soil material is removed as fast as it is weathered.

(iii) *Transport by leaching.* A lot of soil material is also moved from higher to lower parts of slopes by the lateral flow of water within the soil. Most soil constituents transferred by such *throughflow*, including soluble salts, clay, humus, and iron and aluminium sesquioxides, are removed in solution by leaching. On slopes, water moving through permeable materials often reappears as *springs* at junctions with less permeable strata. Around the springs, soils are enriched in bases (e.g. Ca, Mg) by *flushing*.

In summary, soils on lower slopes tend to be not only deeper and wetter than those on the upper slopes but also more enriched by a wide range of leached material. Soils at the top and bottom of slopes thus resemble the A- and B-horizons of individual soils.

4. Soils and time

(a) Soil development and the steady state

(i) *Early development stages.* Soils develop over a long period of time. Immature or 'raw' soils have developed very few properties. They contain little in the way of weathered fragments and organic matter and the horizons are hardly differentiated. Plate 2.7 shows a young podzol 200–300 years old forming at the surface of a soil. The soil horizons are fairly shallow and the grey horizon is only gradually becoming differentiated from the darker A-horizon above it.

As soils continue to evolve, their properties develop rapidly (see Figure 2.17a). Organic remains are added to the soil from growing vegetation. These help to build up supplies of water and nutrients. Weathering processes increase the extent and depth of the weathered B-horizon. In time, soil horizons become increasingly distinct as a result of organic matter being added and translocation taking place.

(ii) *Mature, 'steady-state' condition.* As soils and their properties reach maturity, they begin to stabilise. At this stage any loss of material or energy from the soil (e.g. humus or leaching) is made good by equal inputs of substances (e.g. litter; nutrients brought in by rainfall or released by weathering). As shown in Figure 2.17a, mature soils do not change, or else change only very slowly with time. They are referred to as *equilibrium soils*, i.e. in balance or in 'steady state' with their overall environment.

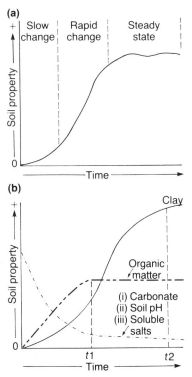

Figure 2.17 Attainment of the 'steady-state' condition in soil development: (a) three-phase theoretical model; (b) variation in selected soil properties with time (Source (b): Birkeland, 1984)

51

Plate 2.7 A thin podzol about 200 years old, with A- and Ea-horizons, lies above a very thick layer of peat. The peat layer has buried an older podzol. Such buried soils and horizons are called 'palaeosols'. Location north Norway. (Photograph: P.Worseley)

Young podzol ≈ 200 — 300 years BP { Ah ——

Ea ——

Palaeosol   Peat ≈ 6000 years BP {

Palaeosol   Podzol ≈ 9000 years BP {

Ea ——

Bs ——

The iron-humus podzol shown in Plate 2.8 may be considered a mature, equilibrium soil. It is located in the same general area as the immature podzol (Plate 2.7) but is between 6000 and 9000 years old – a sufficient time for podzols to attain steady state. The horizons are well marked. A thickish, black, peaty, surface A-horizon lies above a dramatic, massively bleached Ea-zone. Below this layer the Bs-horizon is also well developed in the bright-reddish colours of accumulated ferric iron.

(b)   *Time scales in soil development*

The time taken for any soil to reach maturity depends on environmental conditions, especially climate. For instance, soils will reach maturity quicker in moist, temperate environments (e.g. Britain) than in cold arctic areas where extreme cold slows down the rates of weathering and decay of organic matter, and so hinders soil formation.

Achieving maturity also depends on the nature of the soil property being studied. As shown in Figure 2.17b, the organic and carbonate contents of a soil tend to reach a steady state (t1) before that (t2) of the

Plate 2.8 A mature, 'steady state', iron-humus podzol about 9000 years old with well-developed horizons. Location north Norway. (Photograph: P. Worseley)

O

Ah

Ea

Bh

Bs

Moraine
C

more slowly forming clay minerals. For example, at Glacier Bay in Alaska, young podzols developing on freshly exposed glacial moraines lost most of their surface calcium and accumulated enough concentrations of organic matter to be in steady state within 50 years.

We may also expect that surface A-horizons will develop before sub-surface B-horizons. In addition, while podzols may reach a steady-state, mature condition in as little as 100–200 years, many take more than 1000 years. These time scales are dwarfed by those necessary for the formation of brown soils (10 000–100 000 years) and for some tropical soils (as much as 100 000–1 million years).

(c) Environmental change

In one sense, the idea of the 'steady-state' soil is unrealistic, because its attainment allows only one soil-forming factor to change over time – time itself. In reality all of the soil-forming factors, including climate, vegetation, parent materials and topography, are subject to greater or lesser changes over a period of time. Such changes alter the rate and character of soil development and may prevent equilibrium soils from being achieved. Moreover, some soil properties, which were formed under one set of environmental conditions, may remain in the soil as *relict* features despite subsequent environmental and soil changes. Soils which exhibit such relict characteristics are known as *palaeosols*.

(i) *The Pleistocene glaciations*. From about two million years ago to 10–12 000 years ago, most of Britain was covered in a series of great ice sheets. These removed any pre-existing soil and regolith from the uplands and deposited these materials in lower-lying areas. As a result, bare rock was left in mountain districts, while a great range of glacial deposits was laid down in valleys and lowland plains. The deposits include glacial sands and gravels (from melt-water streams); wind-blown fine silt (loess); periglacial, frost-shattered rock (called coombe or head); colluvium (deposits at the base of slopes). The role of this glacial activity in soil formation is very significant. For instance, when soil maps of Britain are compared with those depicting solid geological formations and those showing superficial glacial deposits (together with river alluvium and recent peat deposits), the soils reflect the latter rather than the former.

(ii) *Buried soils of the post-glacial period*. The post-glacial period covers the time between the last glacial retreat (12 000 BP) and the present day. During the early and middle parts of the post-glacial period, many podzols and brown podzolic soils began to form over bedrock and drift deposits in upland Britain. Many of these soils, as well as the stumps of some of the original trees which grew in them, are now covered in blanket peat. Accumulation of peat in the uplands began during the Atlantic period (a particularly wet phase of the middle part of the post-glacial period) and has been sustained since then by a wet climate and by human activity (see page 81). Such buried soils, or subsoils, are examples of palaeosols. They serve to remind us of the changing environmental history associated with soil formation (see Plate 2.7).

ASSIGNMENTS
1. (a) *Define the concept of the soil catena.*
   (b) *Describe the changes in soils and their properties as shown in Table 2.2.*
   (c) *Using the data in Table 2.2, give three reasons why soils should vary along hill-slopes.*
2. *Refer to Figure 2.18.*
   (a) *Describe and identify soils A–E using the soil types 1–4.*
   (b) *Describe the catenary sequence of soils shown on the sandstone upland.*
   (c) *Using information from the answer to question 1 and the text,*

Table 2.2 Variation of soil properties according to hill-slope, in a humid Arctic environment (northern Norway)

	Soil property	Gentle slope	Main-slope	Foot-slope
1	A-horizon thickness	15 cm	25 cm	30 cm
2	B-horizon thickness	10 cm	30 cm	55 cm
3	Percentage humus in Bh-horizon	2.1	2.2	4.4
4	Percentage sesquioxides (Fe_2O_3 + Al_2O_3) in B-horizon	25.3	29.1	31.2

Source: Birkeland, 1984

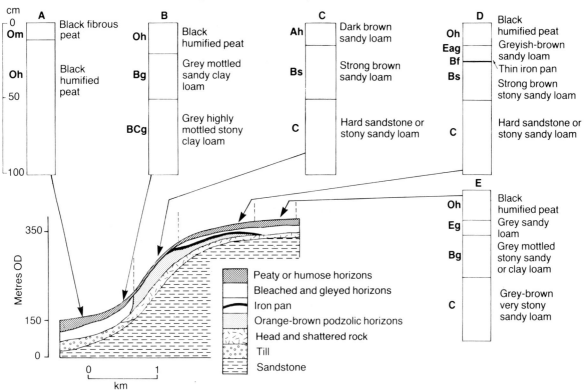

Figure 2.18 Catenary soil sequence on a sandstone upland in Wales (Source: Soil Survey of England and Wales, 1984)

Soil types
1. Peaty iron pan podzol with gleying
2. Brown podzolic soil
3. Peat soil
4. Peaty surface-water gley soils (2 examples)

explain: (i) the distribution of peat soils and peaty horizons; (ii) the distribution of gleyed soils or horizons with respect to slope and iron pan formation; (iii) the peaty podzol–brown podzolic soil transition.

(d) Comment on the problems of using a soil catena in: (i) areas of varied geology and distribution of parent materials; (ii) areas where there is significant human impact along slopes.

(e) For what types of area do you think the soil catena would be useful in surveying and mapping soils?

3. Refer to Figure 2.14.

(a) For each soil-forming factor (except time), give an example of a direct effect that it has on the soil.

(b) For each factor, provide an example of an indirect effect on soil formation.

(c) Explain briefly why interaction between the soil-forming factors makes any precise study of their effects on the soil a difficult task.

Key Ideas

Introduction: Processes and Factors

1. Soil formation takes place when horizons develop within the soil profile.
2. Three main soil processes, i.e. weathering, the incorporation of organic matter, and the movements of water, determine soil formation.
3. Such processes are themselves controlled by a set of environmental or soil-forming factors: climate, parent materials, topography, organisms, and time.

A. Processes of Soil Formation

1. A useful model of soil formation depicts the soil as the end-product of additions, losses, transfers and transformations of materials.
2. Weathering is essential before soil formation can begin. It controls the depth of the soil (B-horizon).
3. Incorporation of organic matter is responsible for the development of O- and A-horizons.
4. The water budget of a soil is the difference between precipitation input and potential losses from evapotranspiration (evaporation + plant transpiration).
5. Where precipitation exceeds potential evapotranspiration in well-drained sites, there is a net downward movement of water through the soil.
6. Waters percolating downwards through the soil cause the transfer or translocation of soil materials from surface to deeper layers.
7. Eluviation is the washing out or removal of any soil material, while illuviation is the washing in and deposition of materials within soils.
8. Waterlogging (the result of poor drainage) produces gleying, the removal and redeposition of iron, and a build-up of peaty organic layers at the surface of soils.
9. Gleying causes the reduction of red (ferric) iron compounds to colourless or grey (ferrous) iron compounds.

B. British Soils

1. The soils of Britain can be grouped into five major classes: brown soils, podzolic soils, gleys, peat soils and 'raw' soils.
2. Brown soils are moderately drained soils found mostly in the warmer, drier lowlands of Britain and are of high agricultural value.
3. They can be sub-divided into typical, leached, acid, and calcareous types.
4. Podzolic soils are strongly leached soils which are widespread in the cooler, wetter uplands of Britain and are of limited agricultural use.
5. Sub-divisions of the podzols include typical, peaty, humus-iron, iron pan, and brown podzolic types.
6. Gleys are poorly drained soils and have characteristic grey, or red-mottled, sub-surface horizons.

7. Groundwater and surface-water gleys can be identified but any soil subject to waterlogging can become gleyed, e.g. brown soils, podzols.
8. Peat or organic soils are the result of severe waterlogging and are so defined when their surface, organic layer is at least 40 cm deep.
9. Rain-fed peats (e.g. raised bog, blanket bog) are highly acidic but valley or basin peats vary in acidity according to the nutrient content of their drainage waters.
10. 'Raw' soils are young, immature soils with a limited formation of horizons, e.g. Alpine soils, sand dune soils, rendzinas on chalk and limestone.

C. Soils and Environment

1. The soil-forming factors act and interact in a complex way to effect soil formation.
2. Because rocks vary in their resistance to weathering, the parent materials influence the rate and depth of soil formation.
3. Parent rocks also affect the nutrient content and permeability of individual soils.
4. The soil catena is a term used to describe the lateral variation of soils on hill-slopes.
5. Soils become differentiated on slopes because of variations in drainage, in the surface transport of sediment and in leaching.
6. Soils develop from 'immature' to 'mature' stages over a period of time, eventually reaching a steady-state condition where they are in equilibrium with their overall environment.
7. The time taken to reach the steady-state condition depends on environmental factors (especially climate) and on the nature of the soil property being studied.
8. The idea of steady-state soils may be unrealistic in view of continuous environmental change.
9. The Pleistocene glaciations have had an enormous effect on the nature and distribution of soils in Britain.
10. Buried soils, or palaeosols, often reveal the changing environmental history behind soil formation.

Figure 2.19 Soil types of the Black Mountain region, Dyfed, Wales

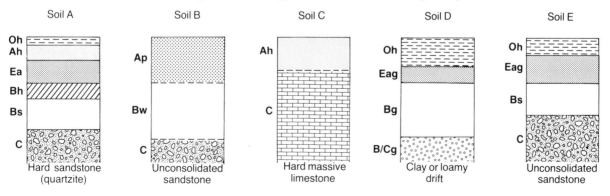

Figure 2.20 A cross-profile of the Black Mountain region, Dyfed, Wales (Source: Soil Survey of England and Wales, 1984)

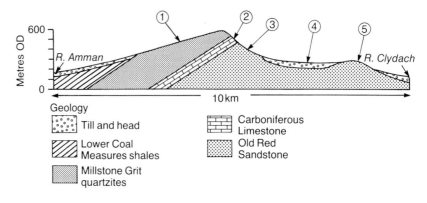

Geology
Till and head
Lower Coal Measures shales
Millstone Grit quartzites
Carboniferous Limestone
Old Red Sandstone

Figure 2.21 Soil regions of Great Britain. At this scale the distribution shown is simplified and highly generalised. Locally there are large variations in soils, so that there may be two or three different soil types within a single field or wood. (Source: Burnham, 1970)

Leached brown soils and related gleys

Acid brown soils and related gleys
Brown podzolic soils and podzols and related gley soils
P Podzols
Blanket peat soils and peaty gleyed podzols
Bare rock and Alpine 'raw' humus soils

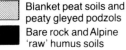

Additional Activities

1. Examine Figures 2.19 and 2.20 and Table 2.3.
    (a) Describe the relief and geology of the Black Mountain region.
    (b) What types of drainage condition and climate are there likely to be at each of the named soil sites 1–5?
    (c) How do your answers compare with the vegetation and land-use data in Table 2.3?

Table 2.3 Vegetation and land-use in relation to soil sites, Black Mountain region, Dyfed, Wales

Soil site	Vegetation and land-use
1	Dry moorland of poor grazing value; coniferous woodland
2	Herb-rich grassland of good grazing value
3	Wet moorland of poor-to-moderate grazing value; some forestry
4	Wet moorland of poor grazing value
5	Cereals, early potatoes, livestock farming

Source: *Soil Survey Bulletin*, 1984

    (d) Describe and name each of the soil types A–E shown in Figure 2.19.
    (e) Identify which of the five soil types A–E is found at sites 1–5.
    (f) Explain your findings.
2. (a) Describe the nature and distribution of the main soil types shown in Figure 2.21.
    (b) Describe the nature and distribution of the main climatic regions shown in Figure 2.22 and Table 2.4.
    (c) Using the data in Table 2.5, describe the water budget and hence the water available for drainage at weather stations A–G.
    (d) What factors other than climate may influence the drainage of a site?

Figure 2.22   Climatic regions of Great Britain (Source: Burnham, 1970)

Warm, dry regime

Warm, wet regime

Cold, dry regime

Cold, wet regime

Very cold, wet regime

▲ A   Weather station

Table 2.4   Climatic regions of Great Britain

Type of climate	Mean annual temperature	Mean annual rainfall
1. Warm dry regime	> 8.5 °C	< 1000 mm
2. Cold dry regime	4.0–8.5 °C	< 1000 mm
3. Warm wet regime	> 8.5 °C	> 1000 mm
4. Cold wet regime	4.0–8.5 °C	> 1000 mm
5. Very cold, wet regime	< 4.0 °C	> 1000 mm

Source: Burnham, 1970

Table 2.5   Mean annual rainfall and potential evapotranspiration at selected British weather stations (see Figure 2.22)

Weather station	Mean annual rainfall (mm)	Mean annual potential evapotranspiration (mm)
A	600	525
B	825	500
C	725	420
D	1200	525
E	1350	430
F	2350	480
G	3100	375

(e)   Examine the relationship between the distribution of British soil types with that of general climatic and hydrological regimes.

3.   Using specific examples, write an essay on the time factor in soil formation.

3 Vegetation Analysis and Survey

A. Description of Vegetation

Introduction

The *flora*, or floristic composition, of an area is the number of different kinds or species of plants which occur within it. The *vegetation* of a region is the overall arrangement and appearance of plants, i.e. the complete plant cover. Vegetation is dependent not only on floristic composition but on the proportions, conditions and distribution of the species present. For instance, two areas may have a similar flora having the same species of grass, shrubs and trees but one area may be dominated by trees, becoming a woodland with an *understorey* of grasses and shrubs, whereas the other may be dominated by its grass species, giving rise to a grassland with isolated trees and shrubs.

1. Composition of vegetation

(a) *Plant communities*

One way of identifying and describing vegetation is to focus on the species which make it up. When vegetation is made up of only one species (e.g. a pine plantation), it is named a *plant society*. Usually, however, vegetation is made up of a particular collection of two or more species and is called a *plant community*.

Plant communities vary according to the types and numbers of species occupying a particular site. Some plant communities, especially those in 'harsh' environments, are comprised of relatively few species and are thus 'species poor'. Shrubby heaths (i.e. heather-dominated communities), which occur in very acidic, lowland environments (see Plate 1.1 on page 13) and in relatively cold, wet upland areas in Britain (see Figure 3.1), may have only a dozen plant species. Other communities, characteristic of less severe environments (or habitats), are relatively 'species rich'. The vegetation of the chalk downland in the drier, warmer parts of south-east England (see Plate 1.2 on page 21) has several hundred species of grass and herb, such as ivy. In parts of the tropical rain forest (see Plate 5.4 on page 117), where there are warm and moist conditions, several thousand species of plant occur.

Figure 3.1 Semi-natural vegetation in Britain

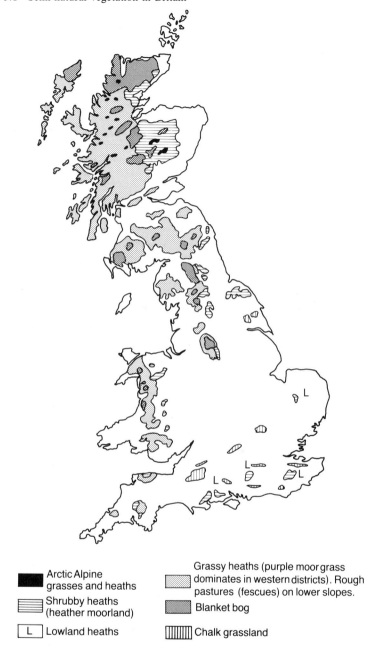

Arctic Alpine grasses and heaths

Shrubby heaths (heather moorland)

L Lowland heaths

Grassy heaths (purple moor grass dominates in western districts). Rough pastures (fescues) on lower slopes.

Blanket bog

Chalk grassland

(b) Plant associations

When a plant community is defined on the basis of its main or dominant species, it is referred to as a *plant association*. For example, a marsh-plant community dominated by sedges is referred to as a 'sedge association'. British woodland communities dominated by beech or oak or ash trees

Figure 3.2   Relationship between plant formations and plant associations (Source: Eyre, 1968)

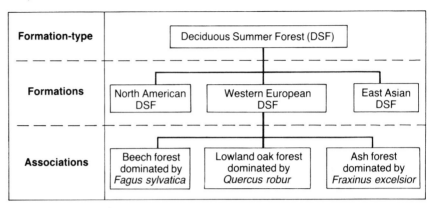

are referred to as beech, oak or ash associations respectively (see Figure 3.2, and Plate 5.2 on page 113).

2.   Structure of vegetation

Plant communities can also be distinguished on the basis of their general appearance and *growth form*, e.g. size, shape, height. In this case we identify woodland or scrub or grassland, without troubling to identify the species present. When plant communities are described by using their overall structure or growth form, we identify what are termed *plant formations*.

Plant formations

(i) *Regional and world-scale.* If we refer again to Figure 3.2, we see that the three woodland associations of lowland Britain form part of a larger vegetation group described by using general growth-form characteristics. Thus, deciduous summer forest of the temperate zone, where trees shed their leaves in the cool winter period, can be identified as the major forest formation of Europe. Likewise this European plant formation can be grouped into an even larger category at the world scale, i.e. that of the deciduous summer forest *formation-type*.

World maps of vegetation distribution (see Figure 5.3 on page 109) tend to emphasise the location of such formation-types. This may be because of the close link between overall growth form and climate, e.g. tropical, evergreen, rain forest with warm, moist environments; temperate, deciduous, summer forest with moisture throughout the year but with a definite, cool period in winter; tundra vegetation (low, woody shrubs and grasses) with cold Arctic conditions.

(ii) *Local-scale: stratification.* Most plant communities have a distinctive vertical structure. Many show vertical layering or *stratification*, i.e. different groups of species occur at different heights above the ground or, in the case of plant roots, at different depths below the soil surface.

Figure 3.3 Woodland stratification in a temperate deciduous forest (Source: Cousens, 1974)

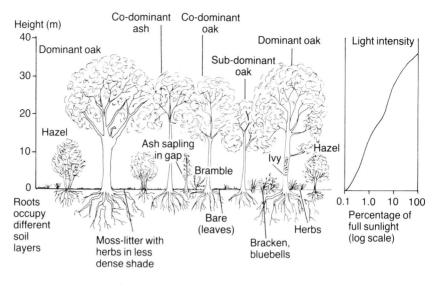

The principal reasons for stratification of vegetation appear to be the availability of, and the competition for, light energy. Between the outer surface of the plant community, where there is full sunlight, and the ground surface, there is a decrease in the intensity of sunlight (see Figure 3.3). Adaptation, competition and selection of plants take place amongst species occupying a given site in order to obtain light energy. As a result, a distinct layering of the plants takes place.

An example of vegetation stratification in a typical summer deciduous forest of the temperate zone is shown in Figure 3.3 and Plate 5.2 on page 113. The main strata include:

(1) *Canopy layer*. The uppermost or dominant stratum consists of the crowns or canopy of the tall, *overstorey* trees, e.g. oak and ash.

(2) *Sub-canopy layer*. A lower layer of smaller trees and/or larger shrubs occurs below the canopy. This layer usually contains younger individuals of the canopy trees (e.g. ash sapling within gaps in the canopy) and/or mature trees and shrubs of smaller species (e.g. hazel) that do not normally reach canopy height.

(3) *Herb or field layer*. A third, lower layer consists of herbaceous (non-woody) plants, such as bracken, bluebells, brambles and ivy.

(4) *Ground layer*. A fourth layer, close to the soil surface, is made up of mosses and occasionally lichens.

The above model of temperate woodland stratification contrasts with the much more complex and varied tropical rain-forest system (see Plate 5.4 and Figure 5.12 on pages 117, 118), where at least five distinct plant layers can be identified.

Comparisons can also be made with non-woodland systems: grasslands often have two, three or more strata (e.g. a tall grass layer, an intermediate low-herb layer, and a ground stratum of mosses); cultivated crops, e.g. fields of wheat or barley, may have only one layer, especially if herbicides are used to remove any competing weeds.

ASSIGNMENTS
1. (a) *Define what is meant by a plant community.*
   (b) *Outline two ways in which plant communities may be characterised and described.*
   (c) *Describe the relationship between plant associations and plant formations shown in Figure 3.2.*
2. (a) *Describe the arrangement of plants shown in Figure 3.3.*
   (b) *Why should plants arrange themselves in such a way?*

B. Development and Distribution of Vegetation

1. A model of vegetation distribution

It is often more difficult to explain the distribution of vegetation, with its internal arrangement of species, than it is to account for the location of, say, drumlins, lakes or cities. This is because vegetation is influenced by many factors (see Figure 3.4), some of which can only be speculated upon as they are either unknown or the product of chance. Nevertheless, four basic considerations help us to understand why a particular type of vegetation is found where it is.

(a) Species arrival

The species which are found in an area of vegetation have to arrive there from elsewhere as seeds, spores, etc. The arrival of flora is influenced by

Figure 3.4 Main environmental factors which affect the development and distribution of vegetation. As in Figure 2.14, there is interaction between factors. For example, climate affects hydrology and soil, while soil influences, and is influenced by, vegetation.

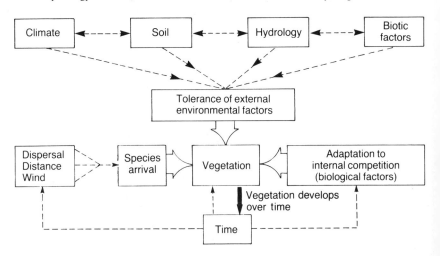

such factors as the ability of the plants to scatter or disperse, the distance travelled, the direction of the wind and the time available for transport.

(b) External environmental factors

The species upon arrival, in order to survive and establish themselves, must adapt to, or tolerate, four different groups of external environmental factors:

(i) *Climatic factors*: e.g. light, temperature, rainfall, humidity, wind, exposure;

(ii) *Soil* or *edaphic factors*: e.g. soil depth and texture, soil chemistry, supply of nutrients;

(iii) *Hydrological factors*: types and distribution of moisture (e.g. ice, liquid water and water vapour) in the atmosphere and the soil;

(iv) *Biotic factors*: e.g. the biotic effects of grazing and trampling by animals, and of burning, cutting and harvesting by humans.

(c) Internal competition

Plant species which are found in an area have to tolerate each other! Existing species have over time adapted themselves to a range of internal, biological pressures from neighbouring species competing for light, space and resources (see *stratification*, page 62).

(d) Time

In any given habitat, vegetation will change over a period of time (see next section), so that 'young' plant communities are quite different from older, mature communities in the same locality.

2. Change in vegetation

(a) Plant succession

(i) *Description of succession.* An area of bare ground, such as freshly exposed rock, or an area cleared of vegetation by fire, flood or human activity, does not remain empty of vegetation for long: an uncultivated field or garden is quickly invaded by 'weeds'; sand dunes are colonised by marram grass and similar specialised, drought-resisting grasses and are thus stabilised (see Plate 3.3 on page 78); the beds of shallow ponds and lakes become covered with algae and water-weeds. These early, invading plants are called the *pioneer species*. They are able to colonise open, exposed sites, free of competition from other species. They do not last for long, however.

In an abandoned arable field, pioneer 'weed' communities will quickly be replaced by grassland which, if left ungrazed for a few years, will be invaded by bushes and converted into scrub. The scrub community may last longer than the herbaceous (weed and grass) plant communities but eventually, as shown in Figure 3.5, it may be colonised and overgrown

Figure 3.5 Plant succession in an abandoned field, under cool temperate conditions. The factors of fire, grazing and cutting are considered to be absent. (Source: Whittaker, 1975)

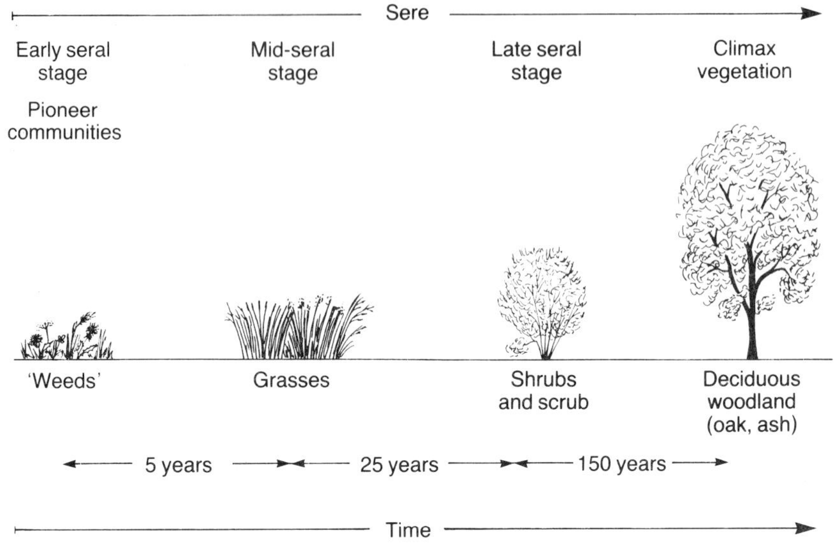

by tree species and so changed to woodland. This process of change in vegetation over a period of time, whereby one group of species replaces another in a given site, is known as *plant* or *vegetation succession*.

(ii) *Trends in succession*. Figure 3.5 suggests that groups of species in a succession do not continue to replace each other indefinitely in any given site. Ultimately, a state of relative stability will be reached where the species exist in equilibrium with their environment. This final plant community (e.g. woodland in Figure 3.5) is referred to as the *climax vegetation*. Notice that the whole process, from the initial establishment of vegetation to the final climax stage, is known as a *sere*. Each group of species in a succession which makes up the sere is called a *seral stage*.

As succession proceeds (e.g. from weeds to woodland), there is usually an increase in: (i) the structural complexity (stratification) and the *biomass* (volume of weight of living plants); (ii) the numbers of species occupying a site; (iii) the growth rate or productivity of the vegetation.

(b) Why does succession occur?

(i) *Internal processes of change*. It is well known that soils often reflect the plants growing on them. The close link between vegetation and soils can be explained by understanding the relationships which govern succession.

The initial establishment of vegetation begins the process of *soil stabilisation*. Pioneer communities are able, by root action, to bind soil particles together and thus to prevent the removal of soil by wind and rain. A developing cover of leafy vegetation also reduces the direct impact of rainfall and restricts wind-speed at the ground surface, thus lessening soil erosion. The protection given to the soil by invading plants is crucial to continuing succession. Without such protection successions would not proceed, and plants would not remain long enough to impress their characteristics on the soil. (See Additional Activities 1 and 2.)

Vegetation also alters the properties of soil by *adding organic matter* which will hold water and nutrients, such as nitrogen, phosphorous and potassium. Plants therefore help to *alter their local environment* or habitat. In so doing, they modify the conditions of the site (soil and atmosphere) to such an extent that other species (which may have been unable to colonise because of lack of water and nutrients or because of exposure), may now find the habitat suitable. Once these later species become established in a succession, they continue to modify their local environment and so encourage further succession.

These progressive environmental changes take place under their own impetus and so may be termed internal or *autogenic* (self-generating).

(ii) *External processes of change.* In contrast to the above changes, external or *allogenic* factors, such as flooding, fire or human activity (e.g. grazing, cutting, ploughing), may disturb vegetation. In this case the original or *primary succession* will be interrupted and what is called a *deflected* or *secondary succession* will begin (see Figure 3.6).

Most successions involve both internal and external factors. For example, in upland Britain the removal of forests by humans has, in wet sites, started the development of allogenic bogs. However, once bogs have formed, autogenic processes become dominant. The growth of the bog-system and the increase of the peat-surface are due to the deposition of organic materials.

Figure 3.6   Types of climax vegetation and successional pathways: (1) primary succession; (2) primary succession interrupted by, for example, soil and biotic factors; (3) retrogressive successsion as a result of disturbance; (4) secondary succession.

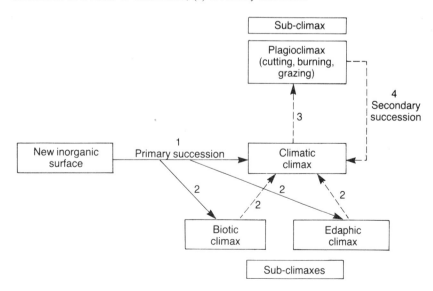

(c) Climax vegetation

(i) *Monoclimax/climatic climax theory*. Ecologists in North America and Europe in the early part of the twentieth century believed that plant succession in any one region tended to end up with a particular vegetation type. The final stage in succession, the *climax vegetation*, had a fixed and predictable composition and was directly related to, and in balance with, its physical environment. The climax vegetation of a region was dependent principally, even exclusively, on the climate and was therefore called the *climatic climax* vegetation (see Figure 3.6). This theory, which related the climax vegetation to the single factor of climate, became known as the *monoclimax concept*.

The idea of the climatic climax is emphasised at global level because climate plays such an important role in determining the overall distribution of vegetation (see Figures 5.2 and 5.3 on pages 108, 109). The same type of equilibrium that exists between climate and vegetation also exists between soils and climate over extensive regions of the world. These equilibrium soils, in balance with climate and vegetation, are known as *zonal soils* (see Figure 5.1 on page 107).

(ii) *Polyclimax theory and sub-climax vegetation*. Today we still recognise that climate exerts a powerful influence on vegetation. For instance, tundra and tropical rain-forest vegetation will not develop in each other's associated climates, and neither will develop in Mediterranean climates. However, we also realise that the detailed nature of climax vegetation, at the local scale, will be determined by a number of factors ranging from climate to some less obvious such as relief and drainage, the nature of the parent materials and the influence of biotic (e.g. animal and human) factors. When vegetation is prevented from progressing to the climatic climax because of some local factor it is called a *sub-climax* (see Figure 3.6).

Sub-climaxes controlled by relief are known as *topoclimaxes*; *hydroclimaxes* reflect the influence of some drainage factor (e.g. impeded drainage). When the dominant influence is soil condition, *soil* or *edaphic climaxes* occur. *Biotic climaxes* or *plagioclimaxes* result respectively when animal or human interference conditions the final, semi-permanent plant community. When the climatic climax vegetation reverts to a previous phase by some internal or external factor, the process of vegetation change is known as a *retrogressive succession* (see Figure 3.6).

The above ideas, relating the climax vegetation to a variety of causes and factors (and not just climate), make up the *polyclimax theory*.

ASSIGNMENTS
1. *Refer to Figure 3.4.*
    (a) *List four separate reasons which account for the distribution of vegetation.*
    (b) *Why is it difficult to be confident about the precise reasons for the distribution of vegetation?*
2. *Refer to Figures 3.5 and 3.6.*
    (a) *Define plant succession.*

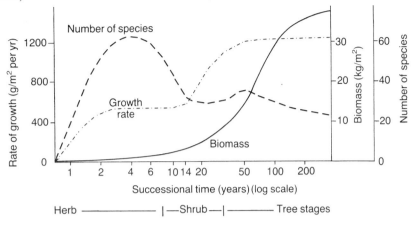

Figure 3.7 Number of species, biomass, and growth rate of vegetation during secondary succession in the Brookhaven oak-pine forest, New York State, USA (Source: Whittaker, 1975)

(b) What is meant by: (i) a sere; (ii) a seral stage?

(c) List: (i) the internal reasons why vegetation should change in a given area; (ii) possible external reasons for vegetation change.

(d) Using your results so far, define primary succession, secondary succession, retrogressive succession.

(e) Define the meaning of climax vegetation.

(f) What factors are likely to determine the climax vegetation?

(g) Describe the relationship between these factors and the type of vegetation climax they produce.

3. Refer to Figure 3.7.

(a) Describe the trends in: (i) vegetation size and complexity (biomass); (ii) plant growth-rate (productivity), (iii) numbers of species found.

(b) Give reasons for your answers.

C. Case Studies: Primary Succession

We shall now look at the development of vegetation in particular types of habitat in Britain. Outlined first are examples of primary successions in both shallow lakes and sand dunes. The next section analyses secondary successions in the British uplands where human disturbance has been a major factor in the establishment of sub-climax communities.

1. Primary successions

Primary successions are started where vegetation has not been present before. *Hydroseres* and *haloseres* are found in shallow fresh and brackish water respectively and are best seen at fresh-water lake margins and at coastal salt marshes. *Psammoseres* occur on sands (mostly sand-dune systems), while *lithoseres* develop on rocks exposed after volcanic eruption or landsliding or on recently exposed glacial deposits.

2. Succession in lakes

The hydrosere at Sweet Mere, near Ellesmere, Shropshire, is a good example of the colonisation of open fresh water by vegetation. As shown in Figure 3.8, Sweet Mere is one of a series of shallow lakes formed in hollows above poorly drained glacial till and fluvioglacial drift deposits. They were created when large ice blocks, which were buried within the drift, melted at the close of the ice age.

The distinctive hydrosere sequence at Sweet Mere is shown in Figure 3.9. Stage one, the pioneer community, begins with rooted but submerged aquatic plants followed by floating, leaved aquatic plants (e.g. water lilies). These species are followed, in succession, by bullrushes (i.e. plants rooted in water but with their flowering parts above), and then by sedges, willow and alder (see Plate 3.1 on page 74). The rise of alder to dominance (stage 5) marks the end of the earlier, more aquatic phase of development. This stage is very clearly seen from the large number of species that help to establish it. At each stage in the succession plants help to stabilise the lake- or pond-bank. They entrap sediment, add organic matter and nutrients to the site, and help to raise the level of the pond-bed.

With the rise of birch to co-dominance (stage 6) and then dominance (stage 7), the ground level rises above the water table. Now the later, more terrestrial phase of the succession begins, where marked changes occur in the composition of soil and species. With the establishment of birch, an acid, peaty surface layer becomes consolidated. The pH of this layer is 3.7–4.3 and contrasts with the pH in water of 7.3 at stages 4 and 5. More than half of the species associated with the wet, alder phase

Figure 3.8 Sweet Mere and other shallow lakes (meres) near Ellesmere, Shropshire

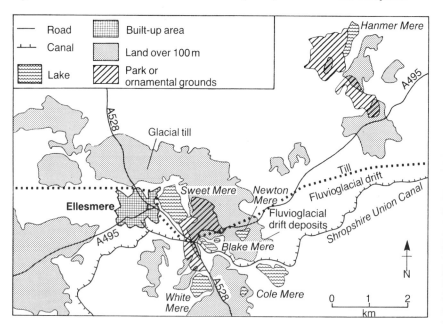

Figure 3.9   Hydrosere at Sweet Mere, Ellesmere, Shropshire (Source: Cousens, 1974)

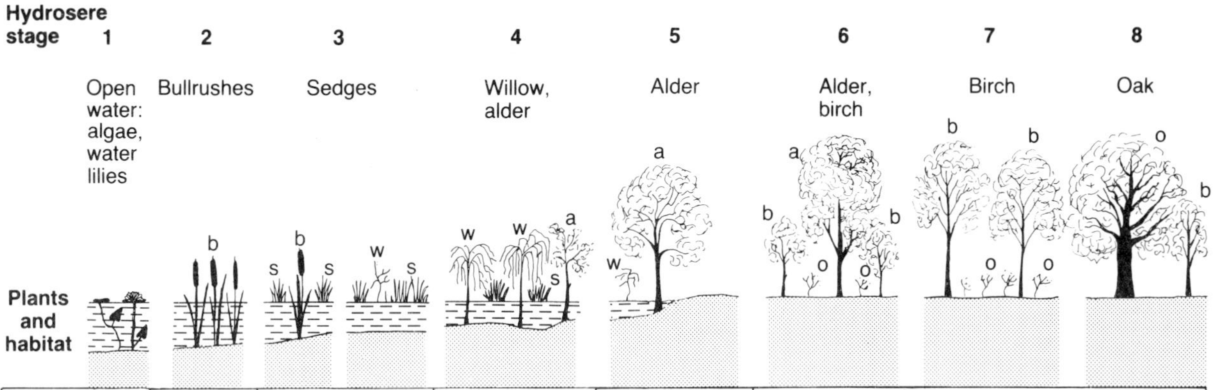

Hydrosere stage	1	2	3	4	5	6	7	8
	Open water: algae, water lilies	Bullrushes	Sedges	Willow, alder	Alder	Alder, birch	Birch	Oak

Habitat description	Reed swamp	Marsh or fen	Open wooded fen	Closed wooded fen	Woodland		
Habitat processes	Accelerated deposition of silt and clay. Floating raft of organic matter forms ⟶ thickens		Raft now a mat resting on mineral soil	Black mineral soil revealed in patches. Earthworms	Ground level now above water table; oak seedlings	Birch canopy forms; oak saplings	Oak grows through and then over the birch
pH level	–	–	7.3		4.3	3.7	–
Number of species of plant	6	10	14	26	18	14	10

(stage 5) disappear to be replaced by a smaller number of new invasions, mainly trees, shrubs and grasses. With the establishment of birch, therefore, the ground becomes completely covered with a field layer of grasses and herbs while most of the large number of distinctively marsh species have disappeared. The last two stages (7 and 8), in which the birch is dominant alone and is then apparently succeeded by oak, are influenced by artificial drainage (allogenic influence). These stages are marked by the local occurrence of bracken in the field layer. The acidity of the surface peat is very high and no varied oakwood flora has developed (i.e. the oakwood is species poor).

3.   Succession in sand dunes

(a)   The survey

In 1985 a soil-vegetation survey was carried out across the large system of sand dunes at Gibraltar Point, Skegness. The survey followed a transect line AB beginning near the sea and ending up about 1000 metres inland (see Plate 3.2 on page 74). Figure 3.10 indicates the location of the 15 recording sites, where soils and vegetation were examined along the transect, in relation to the form and approximate age of the dune system. Areas of young salt marsh, called 'saltings', between the sand dunes, which had their own characteristic soil and vegetation types, were not included in the survey. At each recording site, when vegetation was

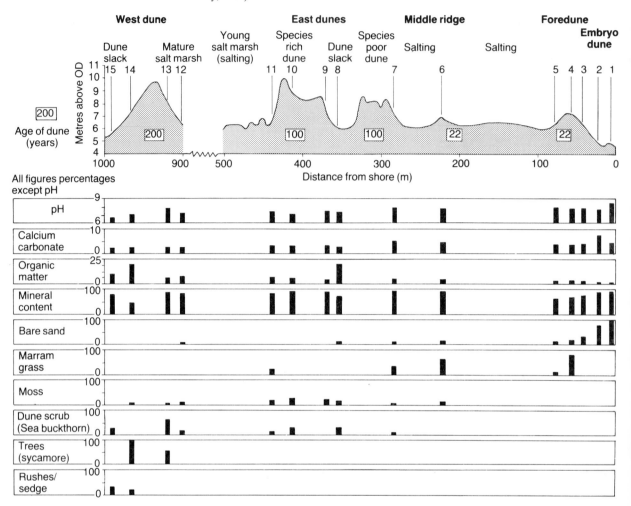

Figure 3.10 Ground area (i.e. percentage cover) of selected indicator species, and physical habitat conditions across sand dunes, Gibraltar Point, Skegness, June–July 1985 (Source: Day, 1986)

present, each species found was sampled as a 'percentage cover' of a $\frac{1}{4}$ m² (i.e. 50 cm-sided) quadrat frame and plotted on a data sheet (see page 88). A selected number of species from the total recorded were chosen to indicate important soil-vegetation relationships across the dunes. These 'indicator' species are shown in Table 3.1. The slope of the land, area of bare sand, soil pH, and contents of calcium carbonate, minerals and organic matter were also measured at each recording station.

(b)   Analysis and interpretation of results

Several types of dune landscape are described on pages 78–9.

Plate 2.1 Brown soil showing a dark surface, mull-humus layer or A-horizon lying above a light brown, weakly-illuviated, mineral B-horizon. Location UK. (Photograph: J. Tivy)

Plate 2.2 Podzol showing a strongly bleached Ea-horizon beneath a black, surface mor-humus layer. A red, iron-rich B-horizon underlies the Ea-layer and is stained black in patches with organic remains. Location USA. (Photograph: J. Tivy)

Plate 2.3 Gleyed brown earth, Derbyshire. A thick, mull-humus A-horizon rests on a clay B-horizon, stained red with compounds of ferric iron. A gleyed clay zone with grey ferrous iron can be seen at the base of the profile. (Photograph: G. O'Hare)

Plate 2.4 Peat soil. This photograph illustrates a very deep, black, surface horizon of poorly decomposed organic material (peat) lying above yellowish-brown mineral soil. Location UK. (Photograph: J. Tivy)

Plate 2.6 Rendzina showing a dark, mull-humus A-horizon developing on calcareous bedrock. Location USA. (Photograph: J. Tivy)

Plate 3.1 Plant succession in lakes and ponds. This photograph shows a wet dune slack at Gibraltar Point, Skegness, being invaded and colonised by plants. The hydrosere sequence is free-floating plants, rushes, sedges, and alder shrubs and trees. (Photograph: D. Day)

A–B transect
1 Embryo dunes
2 Foredunes
3 East dunes
4 West dunes
5 Old salt marsh
6 Young salt marsh
7 Dune slack
8 Agricultural fields

Plate 3.2 Aerial photograph showing transect AB across the dune system at Gibraltar Point, Skegness, Lincolnshire (Photograph: Clyde Surveys, Maidenhead)

Plate 3.5   Seaward-facing slope of the east dunes, Gibraltar Point, Lincolnshire. The dunes here are dominated by marram grass with some rosebay willow herb. (Photograph: D. Day)

Plate 3.6   Old, mature, west dune (seaward-face) at Gibraltar Point, Lincolnshire. Sea buckthorn is the principal plant in this scene, with 'terrestrial' grasses such as *Agrostis* species. (Photograph: D. Day)

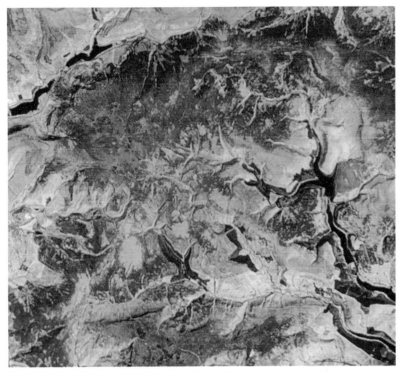

Plate 3.8 Satellite imagery (Landsat 5) of Kinder Scout and Bleaklow, Peak District, April 1984. This false-colour composite distinguishes between peat bog (red), shrubby heaths (dark brown), grassy heaths (pink and light brown) and lowland improved pasture (light green). Coniferous woodland is depicted in dark green while reservoirs are shown in black. (Imagery and photograph: G. O'Hare)

Plate 5.3 Mediterraneal vegetation, southern Spain. This is a patchwork of vegetation with low woody scrub (garrigue), taller shrubland (maquis) and, in the background, planted woodland of Corsican and Aleppo pine (Photograph: G. O'Hare)

Plate 5.5 Tropical latosol. A red, deeply weathered, iron-rich B-horizon is the main feature of this soil. Lying above the B-layer is a darker, surface, A-horizon with some incorporations of humus. Location USA. (Photograph: J. Tivy)

Plate 5.6 Red-yellow latosol. This soil resembles the tropical latosol and has a red, sesquioxide-rich B-horizon. Being formed under cooler sub-tropical conditions, silica is not so easily removed, however. As a result, a bleached, silica-rich Ea-horizon is often seen, which means that there is podzolisation. Location USA. (Photograph: J. Tivy)

Plate 5.9 Chernozem or grassland soil. A deep, black, mull-humus A-horizon lies above a C-horizon of calcareous, wind-blown loess. Nodules of calcium carbonate can be seen in the C-horizon. Dark channels indicate the presence of animal burrows. Location USA. (Photograph: J. Tivy)

76

Table 3.1 Vegetation analysis (percentage cover) of selected indicator species across sand dunes, Gibraltar Point, Skegness (June–July, 1985)

Habitat (site number)	Bare sand	Couch grass	Lyme grass	Marram grass	Red fescue	Moss species	Rosebay willow herb	Brambles	Sea buckthorn	Yorkshire fog grass	Sycamore	Rushes, sedges
Embryo dune (1)	99											
Foredune (2)	73	24										
Foredune (3)	38	8	33									
Foredune (4)	12	–	5	79								
Foredune (5)	8	–	23	25	30							
Middle ridge (6)	10	1	–	60	10	5						
East dune (7)	5	6	–	25	5	2	32	6	5			
*East dune (8)	3					12	28	17	23			
East dune (9)						20	25	38	27			
East dune (10)						24	25	12	13			
East dune (11)		3		21	1	17		9	12			
West dune (12)	3				5	5		15	60	2		
West dune (13)						3		28		1	50(c)†	
*West dune (14)						1				3	100(c)	14
*West dune (15)						1			27	2		27

* Dune slack areas   (c) represents canopy cover of trees

Source: Day, 1986

Plate 3.3 Embryo dunes forming around, and to the lee of, pioneer vegetation of sea couch-grass, Gibraltar Point, Lincolnshire (Photograph: D. Day)

(i) *Embryo dunes*. These are the youngest, lowest and most scattered dunes of the system (see Plate 3.3). Being nearest the sea and low lying, these dunes are prone to easy removal by easterly winds and storms, and are often splashed by sea water. This is the area to be first colonised by plants. However, it is a fairly hostile, saline environment where pH levels are 8–9. Very few species can tolerate such conditions and 99% of the dune is bare sand. One of the most notable species found here is sea couch grass, where individual, scattered plants can be seen to entrap sand, stabilise the dune system and so encourage further growth of the dune around their roots and stems.

Plate 3.4 Foredunes at Gibraltar Point showing lyme grass on the lower, seaward-facing slopes and marram grass dominating the middle and upper slopes and ridges. (Photograph: D. Day)

(ii) *Foredunes*. Twenty metres or so landward of the embryo dunes are the older, higher but still scattered (see Plate 3.4) foredunes. They are more highly vegetated than the embryo dunes with the percentage of bare sand falling from around 73% on the seaward dune slope to only 8% on the landward. Couch grass increases its percentage cover on the seaward slopes while lyme grass, another drought-resistant and salt-tolerant grass species, invades and establishes itself on the lower landward and seaward margins.

The most noticeable species of this dune, however, and indeed of the middle ridge and parts of the seaward slope of the east dunes, is marram grass. This species is able to tolerate a dry, mobile habitat of shifting sands. Once the marram roots its horizontal stems or *rhizomes* through the sand, it binds the sand together enabling a small number of other species (e.g. red fescue grass) to invade on the more protected, landward slope.

(iii) *East dunes*. About 300–400 metres from the sea lies a major series of dunes called the east dunes. This system is highly vegetated with marram grass, brambles and rosebay willow herb (see Plate 3.5 on page 75). The increased age and stability of this system is indicated by the appearance and gradual increase of ground moss and scrub (sea buckthorn) across the system. Indeed, the highest number of species in the whole dune complex at Gibraltar Point was recorded at sites 9, 10 and 11 in the western section of the east dunes. The maturity and stability of this dune, compared with the foredunes and embryo dunes, can also be seen in the reduction of soil pH and the content of calcium carbonate and in the build-up of soil organic matter (and therefore soil moisture) which accompanies longer-term colonisation by vegetation.

(iv) *West dunes*. The most landward dune, the west dune, is the oldest, most protected and most stable dune. As a result, sea buckthorn increases in dominance, especially on the seaward slope, and 'terrestrial' grasses, e.g. *Agrostis* species, Yorkshire Fog grass (see Plate 3.6 on page 75), and herbs appear. Most typical of this dune, as shown in Plate 3.7, is the establishment of a dense tree cover, in particular of sycamore, on the more protected westward-facing slope.

(v) *Dune slacks*. Between some of the dune ridges are low-lying depressions, the dune slacks. In summer these low-lying areas have their surface just above the fresh-water table but from October to April they are often covered by a rising water-level which submerges all the plants in those areas. Such conditions encourage many plants to thrive, especially sea buckthorn in the east dune slack and rushes, sedges and alder trees in the west dune slack (see Plate 3.1).

ASSIGNMENTS
1. *Refer to Figures 3.8 and 3.9.*
   (*a*) *Explain the origin and distribution of the shallow-water lakes.*
   (*b*) *Describe the changes in vegetation from open water to dry land across Sweet Mere.*

Plate 3.7 Well-established sycamore woodland with a rich grass and shrub understorey can be found in moist (but not flooded) depressions in the west dune, Gibraltar Point, Lincolnshire. (Photograph: D. Day)

(c) *What factors account for these changes in vegetation?*
(d) *Account for the variation in the number (variety) of species with reference to: (i) shade conditions; (ii) soil pH; (iii) moisture content of the habitat.*
(e) *Describe the sequence of vegetation remains and sediment which is likely to be found in the soil* below *the oak forest.*

2. *Refer to Figure 3.10 and Table 3.1.*
   (a) *Describe the sequence of dunes shown.*
   (b) *Describe the changes in vegetation across the dune system.*
   (c) *Account for the changes you have mentioned by using the habitat data shown in Figure 3.10.*

D. Case Studies: Secondary Succession

The vegetation of Britain is almost everywhere subject to a great deal of human interference, either directly (e.g. by cutting, removal of forests) or indirectly (by fire or domestic grazing animals). Before going on to look at the types of plant communities which human disturbance creates in particular habitats, what do you think would be the natural climax vegetation of Britain in the absence of human actions?

1. Natural climax vegetation

In most parts of lowland Britain and in many upland areas, the natural (climatic) climax vegetation would appear to be forest. This forest would generally be dominated by oak (see Plate 5.2 on page 113) but, under certain soil conditions (e.g. on chalk/limestone), by beech or ash. In central Scotland, pine formerly dominated the forest until widespread felling took place in the seventeenth and eighteenth centuries. In poorly drained areas in the west of Britain and in mountainous regions, the growth of blanket bog, made up mostly of bog moss and cotton grass, may naturally have replaced forest in high P/Pet conditions (e.g. of high rainfall and low evaporation). Such blanket bog would represent another kind of climax vegetation. Above the tree line, sub-Alpine scrub and Alpine heaths and grasses would predominate and form a third type of natural climax vegetation (see Figure 3.1).

2.   Plagioclimax communities of the uplands

The changes in zones of vegetation according to altitude have been much obscured in Britain by human management. As shown in Figure 3.1, the removal of forests has encouraged the further spread of blanket bog (see page 54). Sheep-grazing and fire have prevented the natural regeneration of forest, already near the limits of its range in the uplands. Today the highest oak woodlands in Britain lie at less than 450 m OD on Dartmoor and in the Lake District. Fragments of the hardier, native Scots pine forest occur up to 640 m OD in the Cairngorms.

Most of the upland area of Britain, which was once forested, is now made up of one of the following broad categories of vegetation/land-use types. These categories are arranged in order of decreasing grazing quality and/or human management. Their general distribution in the Peak District is dramatically highlighted from space in Plate 3.8 on page 76.

(a)   Improved pastures

These contain introduced species of high agricultural value, e.g. rye-grass, cock's foot and white clover. They are most usual on the lower slopes of the uplands (150–200 metres) where soils are mainly of the brown soil type with pH values above 5.2 and often above 6.0. Rainfall is usually less than 1000 mm.

(b)   Rough pastures

These are often located on the higher, wetter slopes immediately above the improved pasture (see Plate 3.9). They are subject to less intense management than the improved pastures and may result from either deterioriation of former improved grassland or partial upgrading of moor-lands (see below). Rough pastures are characterised by the prominence of species such as bent grass and sheep's fescue, rushes and other coarser species. Bracken is characteristic on slightly deeper soils on well-drained slopes. Lower fertility, with moderately acid brown podzolic soils and acid brown soils (pH 4.7–6.0), is typical of rough pastures, and poor drainage is frequent.

Plate 3.9 A mosaic of different vegetation and soil types (or ecosystems), Bransdale, North Yorkshire. Enclosed fields with improved pasture occupy the valley floor and the middle slopes. The upper-valley slopes carry rough pastures and grassy heaths. Above these, a ridge has shrubby heath vegetation. A conifer plantation is found on the left middle-slope, while a remnant of deciduous woodland remains on the valley floor. Typical podzols and peaty podzols are associated with the shrubby heath of the upper sandstone ridge. Brown soils are found in relation to the well-drained lower slopes, but they become gleyed in the poorly drained valley bottoms. Brown podzolic and some peaty podzols are associated with the rough grazings and wet grassy heaths respectively. (Photograph: Institute of Terrestial Ecology/NERC)

(c) Grassy heaths (grass moorland)

Rough pasture grades into grassy heaths in higher or wetter areas. They are typified by the dominance of coarse grasses, such as moor mat grass and wavy hair grass, in moderately wet sites up to 800–900 metres in altitude. Purple moor grass is often found in lower, wetter areas below 500 metres in western Britain, on slopes which receive flushes of acid, peaty water from above. Rushes and bracken may also be present locally. Bilberry and heath bedstraw are quite common plants of these grass moorlands. Soil acidity covers a wide range but many sites, typically on peaty podzols, have pH levels below 4.7.

(d) Shrubby heaths (heather moorland)

Shrubby heaths are so called because of the presence of dwarf shrubs, such as heathers, bilberry and crowberry, as well as gorse, rushes and sedges. One type is found on dry, excessively drained, lowland sites including the fluvioglacial sands of the Breckland, Norfolk, the Bagshot sands of Hampshire and the Weald sandstone (see Plate 1.1 on page

13). The shrubby heaths of the uplands prefer generally drier conditions to those of the grass moorlands and blanket bog. As shown in Figure 3.1, they are mainly located on the drier, eastern flanks of the western mountain ranges, and in upland areas to the east (e.g. the North Yorkshire Moors, Plate 3.9). Shrubby heaths occur on well-drained, very acid podzolic soils, with pH values less than 4.2.

(e) Blanket bog

Such plant communities are found in wet, very poorly drained (i.e. flatter), upland areas. Heather is often present but the distinctive species are sphagnum or bog moss and cotton grass. As with the shrubby heaths, acidity levels tend to be very high, commonly with pH levels around 4.0.

3. Shrubby heaths (heather moorland)

Shrubby heath (heather moorland) is today one of the major components of the vegetation of Britain, with extensive distributions in upland areas (see Figure 3.1 and Plates 3.8 and 3.9). However, during the primary succession and revegetation of these islands after the retreat of the Pleistocene glaciations, heather moorland vegetation would not have been a major representative of any seral stage. This is because it would have been excluded fairly quickly by later stages in succession e.g. by scrub and trees (see Figure 3.11). It owes its present extent and dominant ecological status to human interference.

For instance, over several millenia the clearance by humans of upland forests (i.e. mixed oak forest in England and pine and birch in Scotland) has allowed heather to take advantage of newly created, open, exposed and often nutrient-deficient sites. Upland habitats have lost nutrients as a result of the destruction of forests and the breakdown of the forest-soil

Figure 3.11 (a) Possible pathway of primary succession during the post-glacial period in the low uplands of Britain; (b) subsequent modification of vegetation by human activity produces a variety of plagioclimaxes depending on local conditions.

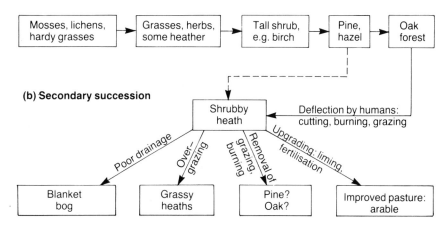

(a) Primary succession

(b) Secondary succession

83

nutrient cycle (see Chapter 4). As trees are removed, they cannot continue to return soil nutrients from deeper layers to the soil surface, and so there is very little in the soil to hold nutrients. As a result, the nutrients tend to be washed or leached away in the drainage waters.

The effects or impacts of subsequent heavy grazing and frequent burning have prevented the regeneration of forests and have allowed heather communities, which are adapted to these impacts, to maintain their present cover. These heather communities will remain for only as long as the forces which maintain them continue. If the maintaining factors of burning and grazing are removed, heather-dominated plagio-climax vegetation will often be quickly colonised by birch and pine and, if conditions are not too wet or acid, possibly by oak. These climax tree species become established because they are better adapted to the new environment, free from grazing and fire.

The patterns of replacement of one community by another in the uplands are rather complex, however, and depend on a variety of local site conditions as well as burning and grazing regimes. As indicated in Figure 3.12, moderately acid, upland rough pastures (e.g. bent/fescue grassland) may be colonised first by bracken, especially if cattle are removed, and then by scrub and woodland, in the absence of grazing. Heather may also colonise the bent/fescue pastures, in the absence of bracken, with reduced pressure from grazing. If grazing continues to the reduced, heather may be invaded by scrub and woodland. However, with

Figure 3.12 Changes in upland grassland and heather communities in Britain as a result of variation in the intensity of grazing (Source: Institute of Terrestrial Ecology, 1978)

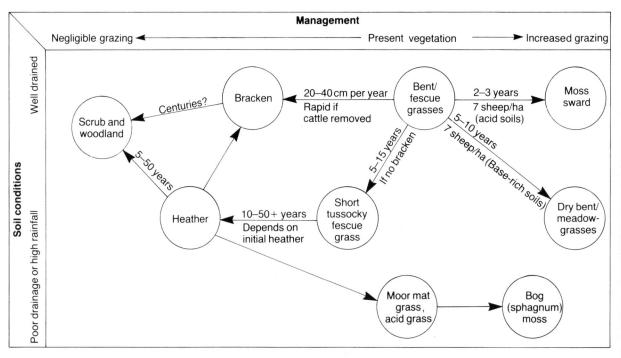

an increase of pressure from grazing, the heather may be replaced by bracken or by moor mat grass and eventually by moss on wet, acid soils.

ASSIGNMENTS
1. (a) *Describe three types of climatic climax vegetation in Britain.*
   (b) *How has human activity interfered with the climatic climax vegetation?*
   (c) *Refer to Figure 3.1 and Plates 3.8 and 3.9. Describe the nature and distribution of five different plagioclimax communities in the uplands.*
2. *The path of secondary succession in the uplands is complex and varied. Discuss this statement, using Figures 3.11 and 3.12.*

E. Field Survey of Vegetation

1. General principles

Most areas of vegetation are usually quite extensive and consist of a wide variety of species which require an expert botanist to identify them. Plant communities need to be simplified, therefore, so that they can be measured and recorded in the field by geographers. There are two basic ways of doing this: to focus on the character of the vegetation itself and to concentrate on the actual area to be surveyed.

(a) *Simplifying vegetation*

One useful method is to concentrate on the geometry and growth form of plant communities. We might study plant-covered areas layer by layer and examine how such stratification provides an effective use of sunlight. Woodland is an ideal example. Using a light meter from a camera, it is possible to consider the availability of light at different heights, seasons etc, within different types of woodland, e.g. deciduous and coniferous forest. The response of plants, especially the ground flora, to light conditions can be studied.

Another method is to group plants into general plant communities. These can be described by identifying several dominant species together with overall growth-form characteristics, e.g. identifying heather moorland with bilberry or purple moor grassland or sphagnum/cotton grass blanket bog. It is also useful simply to count the numbers of species (without naming them) in each vegetation type, to give an idea of the variety of species.

General descriptions of plant communities, however, may be self-defeating. In some instances (e.g. in the survey of vegetation across a sand dune, page 71), it may be necessary to identify and name key indicator-plants and relate them to specific habitats. Common plants can be identified by using well-illustrated texts such as the *New Concise British Flora* by W. Keble-Martin (1982), *Wild Flower Key: British Isles and North West Europe* by F. Rose (1981) and *Collins' Guide to the Grasses, Sedges and Ferns of Britain and Northern Europe* by R. and A.

Fitter (1984). Specimens collected in the field can be taken to specialists and natural history societies. But it must be emphasised that it is now illegal to pick and uproot any wild plant unless permission has been given by the owner or occupier of the land. Certain species are totally protected and their removal is illegal!

(b) Simplifying areas

To separate and identify every plant within a vegetation type would be impossible; plant communities are too extensive and complex for that. We need, therefore, to select or *sample* a small area of vegetation and use this to represent the larger area or plant community.

(i) *Quadrat analysis*. A square frame or quadrat, made of wood or metal, with sides of 50 cm or 1 metre and with cross-strings every 5 or 10 cm, is often used to measure the characteristics of vegetation (see next section). For rich, herbaceous vegetation, e.g. that found on the chalk downlands, a 50 cm-sided quadrat may be suitable, while for larger, shrubby vegetation, e.g. tall heather moorland, a quadrat 1 metre square is preferable.

In measuring vegetation, quadrat frames can be positioned either randomly (i.e. thrown over the shoulder) or systematically (i.e. at regular intervals along a line or over a grid). The *intensity* of sampling, i.e. the number of quadrats recorded, is important in determining the quality of the results obtained: the more quadrats used, the higher the accuracy of the results. When surveying a moorland or a sand dune, about 20–30 quadrats may be used to give reasonably representative results of the whole.

Plate 3.10   Field survey of vegetation near Fort William, Scotland. Students are using 1-metre-square quadrats along a line transect to record the characteristics of a bracken-covered slope. (Photograph: G. O'Hare)

(ii) *Transects*. A *line transect*, e.g. across a feature such as a footpath or from the top to the bottom of a slope, records lateral variation in plant distribution. All plant species (or life forms) touching the line or those touching only at regular intervals (e.g. every 5–10 metres) may be recorded (see Plate 3.10). A *belt transect* records all plants within a suitable distance from the line on one or both sides, continuously or at suitable intervals (e.g. using 1-metre-square quadrats every 5–10 metres). A *profile transect* (see page 72) has the added advantage of indicating topography along the line, by using a level or clinometer to record slope and altitude.

(iii)   *Preliminary study*. This is useful in allowing us to select the test area and methodology. As with all sampling, the *apparent* distribution of plants is a good guide to the choice of method. Gradual or marked variations of vegetation along slopes can be sampled by using a profile transect. If different plant communities in different areas are to be surveyed and compared, then it is better to begin by identifying *uniform* or *homogeneous units* of vegetation. Their respective characters can then be sampled by using quadrats.

2. Recording vegetation characteristics

(*a*) *Quantitative description*

We can record the amounts of plants in an area in one of three ways: by frequency, density or cover.

(i) *Frequency*. This is a measure of the *distribution* of a species (or life form) in an area. It is expressed as the percentage of sample plots (quadrats) in which the species occurs. In each quadrat the name of each species (or life form) is recorded, as shown in Table 3.2. From the data, a table of species in decreasing order of frequency can be drawn up for the vegetation area. In order to assess the relative importance (not just presence/absence) of different species within a plant community, the numbers and/or cover of different species occurring within quadrats must be recorded.

(ii) *Density*. This is the number of individuals of a species (say *A*) per unit area or quadrat. It is calculated as shown:

$$\text{Density of Species } A = \frac{\text{Total number of individuals of Species } A \text{ recorded}}{\text{Number of quadrats recorded}}$$

Table 3.2   Recording the frequency of plant species in an area. A tick means that a species is present.

Species	Quadrat number							(n)	Percentage frequency
	1	2	3	4	5	6	7		
A	√	√					√		Number of plots in
B	√	√	√	√	√	√	√		which species occurs
C					√				―――――――――― × 100
D	√	√	√			√	√		Total number (n)
E	√				√	√			of quadrats

Figure 3.13 Estimating percentage cover of vegetation by eye. This simple technique is not without its difficulties. For instance, the purple moor grass has an apparent ground cover of about 60%, i.e. distance AB in relation to MN, but the open spaces between the grass leaves will reduce this cover value. Also, plants at different heights (e.g. moss and bilberry below the taller grass and heather) will be under-represented. Can you think of a technique to record vegetation at different levels?

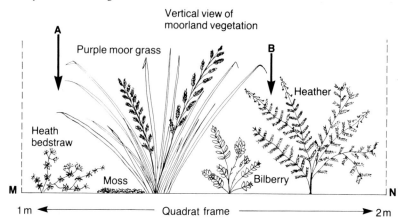

(iii) *Cover*. Density measurements, although useful, are time consuming. A common alternative technique of measuring the relative importance or *dominance* of a species is to assess its cover value in a given area. This is a measure of the amount of ground covered by a species. It is usually expressed as the area covered by the 'crown', stem or mass of a particular plant (see Figure 3.13). A common procedure is simply to estimate by eye the percentage of ground covered in a quadrat by a species. Using a simple scale (see Table 3.3), each species can be measured and given a cover class. The average percentage cover for each species is calculated as shown in Table 3.4. Species can then be arranged in decreasing order of cover.

Table 3.3 Simple cover scale

Class or scale	Percentage range of cover
X	< 1%
1	1–10%
2	11–25%
3	26–50%
4	51–75%
5	76–100%

Table 3.4 Recording mean cover value

Species	\>Quadrat number 1	2	3	4	5	6	7	(n)	Mean cover class
A	4	5	3	3	4	5	4		
B		3	3					1	Total of classes (i.e. 4 + 5 + etc)
C					1				
D	3	X	2				X		Total number (n) of quadrats
E			2	4					

Table 3.5 The Braun-Blanquet system of rating plant abundance. For example, plants may have grouped or tufted appearance (Soc 2) and cover a little (e.g. Cover 2) or a lot (e.g. Cover 5) of any quadrat.

Cover	Grouping (Soc = Society)
+ = sparse, cover small	Soc 1 = isolated individuals
1 = plentiful, but cover small	Soc 2 = grouped or tufted
2 = numerous, cover ≥ 1/20	Soc 3 = patches or cushions
3 = any number, cover $\frac{1}{4}$–$\frac{1}{2}$	Soc 4 = colonies or carpets
4 = any number, cover $\frac{1}{2}$–$\frac{1}{3}$	Soc 5 = pure populations
5 = covering > $\frac{3}{4}$ of the area	

(b) *Qualitative description*

It may be enough to obtain data in qualitative form. This can be done in terms of the relative abundance and grouping of species. A good rating scale for this purpose is the Braun-Blanquet system (see Table 3.5).

Key Ideas

A. Description of Vegetation

1. The vegetation of an area is the complete plant cover.
2. Vegetation can be characterised on the basis of the composition of its species and/or its physical form or structure.
3. Plant communities are units of vegetation with two or more species.
4. When plant communities are defined in terms of their main or dominant species, they are called plant associations.
5. A plant community characterised by general form or structure is referred to as a plant formation.
6. A notable structural feature of plant communities is their vertical layering or stratification.
7. Four distinct layers can often be identified in a temperate deciduous woodland: canopy, sub-canopy, field and ground strata.

B. Development and Distribution of Vegetation

1. A useful model explaining vegetation distribution considers species arrival, the influence on species of internal and external habitat factors, and plant development through time.
2. External factors influencing vegetation include climate, soil, hydrology, and the effects of people and animals.
3. Internal factors, such as the competition among plants for space, light and resources, also shape vegetation development.
4. The process of change in vegetation over a period of time, whereby one group of species replaces another in a given area, is known as plant succession.
5. With succession there is often an increase in the stability of the plant community, together with an increase in (i) its size and complexity; (ii) its growth rate; (iii) the number of species occupying it.
6. Successions occur because internal or autogenic processes, such as soil stabilisation and other modifications of site and soil by plants, encourage one group of species to replace another.
7. External or allogenic factors, such as drought or flood or cultivation, also cause succession by disturbing the direction of changes in vegetation.
8. The relatively stable end-stage in plant succession is known as the climax vegetation.
9. When the single factor of climate determines the end-stage, a monoclimax or climatic climax vegetation results.
10. When the apparent end-stage is controlled by some local factor or

factors, e.g. soil, relief or humans, it is known as a sub-climax or polyclimax vegetation.

11. Examples of semi-permanent climax vegetation controlled and managed by humans are known as plagioclimaxes.

C. Case Studies: Primary Succession

1. Primary successions take place on new, inorganic (i.e. previously unvegetated) sites.
2. The hydrosere at Sweet Mere, Ellesmere, Shropshire, provides a good example of the colonisation of open fresh water by vegetation.
3. The sequence of vegetation colonisation at Sweetmere is aquatic plants – bullrushes – sedges – willow – alder – birch and then oak.
4. At each seral stage, sediments are entrapped and organic matter and nutrients are added, all of which help to raise the level of the pond-bed and encourage succession.
5. At Gibraltar Point, Skegness, there is a well-defined series of dunes, from the seaward margin inland: embryo dunes, foredunes, east and west dunes.
6. With plant colonisation and succession, these dunes are progressively stabilised, being enriched with organic matter and hence with soil moisture, and having reduced levels of calcium carbonate and pH.
7. In very general terms, the succession moving inland across the dunes is: invasion by grasses and herbs followed by scrub and then by woodland.

D. Case Studies: Secondary Succession

1. Those successions which take place in sites which have been previously vegetated are known as secondary successions.
2. As a result of forest clearances and other effects, humans sustain five types of plagioclimax communities in the uplands.
3. These are, in decreasing order of grazing quality and/or degree of management: improved pastures, rough pastures, grassy heaths, shrubby heaths, and blanket bog.
4. Secondary successions in the uplands can be complex, depending on local environmental conditions and on the degree and type of vegetation disturbance.
5. Shrubby heaths (heather moorland), for instance, may be upgraded and used for cultivation or they may pass to grassland, bog or revert to forest.

E. Field Survey of Vegetation

1. Areas of vegetation need to be simplified before they can be measured and recorded in the field.
2. Vegetation itself can be simplified by concentrating on its physical structure: counting the number of species or grouping species into general plant communities.

Figure 3.14 Simple diagram of sand dune system at Cooloola, Queensland, eastern Australia (Source: Walker *et al*, 1981)

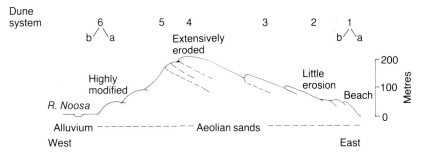

Table 3.6 The approximate age in years of the different dune systems at Cooloola, Queensland, eastern Australia

1	300–500
2	3000
3	6000
4	15–20 000
5	110–140 000
6	over 400 000

Source: Walker *et al*, 1981

Figure 3.15 Thickness of soil horizons and development of soil depth on the Cooloola dune system, eastern Australia (Source: Walker *et al*, 1981)

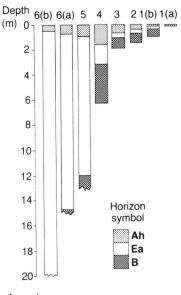

Annual rainfall = 1500 mm
Sub-tropical climate

3. Sampling techniques, e.g. using quadrats and transects, provide a way of simplifying and reducing the area of vegetation to be studied.
4. Quantitative assessments of vegetation can be made by measuring the frequency, density and cover of plant species.
5. Frequency is a measure of the distribution of a species, being expressed as the percentage of sample plots (quadrats) in which a given species occurs.
6. Density is the number of individuals of each species per unit area.
7. Cover is a measure of dominance of a species and is expressed as the area of ground covered by the 'crown', stem or mass of a particular plant.
8. The Braun-Blanquet cover-abundance scale is useful in estimating vegetation qualitatively, i.e. in terms of its relative abundance and grouping characteristics.

Additional Activities

1. (a) Describe the pattern of dune development in Figure 3.14 and Table 3.6.
   (b) Using Table 3.7, describe the variation in plant succession across the dunes, noting: (i) the height and structure; (ii) the biomass; (iii) the species composition of the vegetation.
   (c) Compare your answers with the sequence of change outlined in the section *Plant succession* (pages 65–7).
   (d) Refer to Figure 3.15. Describe the soil changes associated with the different dunes and vegetation systems.
   (e) Account for your answers, noting: (i) the relationship between vegetation and depth of organic matter (A-horizon); (ii) the effects of leaching over a period of time.
   (f) Name the soil types involved.
   (g) Using both soil and vegetation evidence, justify the claim that plant succession here shows both progressive and retrogressive trends.
2. This activity describes the effects on heather moorland as it is colonised by birch woodland.

Table 3.7 Structure of vegetation, biomass and variety of species for each dune system at Cooloola, Queensland, eastern Australia

Dune system number	General description of vegetation (height in metres)	Percentage cover	Biomass index i.e. height of vegetation (m) × percentage cover	Number of species Trees	Total
1a	Dwarf (5)	0.4	2	2	6
1b	Low (12)	46.5	558	5	43
2	Tall (20)	39	780	8	52
3	Very tall, layered (35)	40	1400	8	63
4	Extremely tall, layered (28)	82	2300	9	75
5	Tall, layered (15)	60	900	7	80
6a	Low (9)	27.3	246	3	64
6b	Dwarf (1)	100	100	2	47

Source: Walker *et al*, 1981

Refer to Figure 3.16 and Table 3.8.

(a) Describe the changes in (i) the percentage cover of species and (ii) the number of species in each community, as birch colonises heather moorland.

(b) Describe the effects of birch colonisation on the heather moorland in terms of pH level and the nutrient content of the soil.

(c) What effect does a changing pH level have on: (i) soil organisms; (ii) the breakdown of organic matter; (iii) variety (number) of species?

(d) List the evidence which suggests that, with the invasion of birch woodland, typical podzols under heather change to brown podzolic soils.

Figure 3.16 Mean percentage cover of different species in heather moorland and in adjoining birch communities of different ages, near Advie, Morayshire (Source: Miles, 1981)

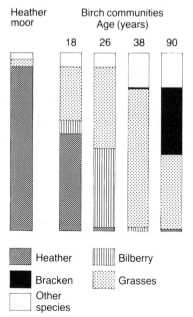

Heather moor   Birch communities Age (years)

18   26   38   90

Heather · · · Bilberry
Bracken · · · Grasses
Other species

Table 3.8 Soil and vegetation characteristics of heather moorland and adjoining communities of birch (*Betula pendula*) woodland. The stands of birch are aged 18–90 years. Advie, Morayshire, Scotland

Surface soil condition	Heather moorland (*Calluna vulgaris*)	Birch woodland colonisation of heather moorland: age of birch communities in years			
		18 years	26 years	38 years	90 years
pH	3.8	3.9	4.0	4.7	4.9
Calcium (mg/dm⁻³)	196	201	207	489	319
Number of earthworms per m³	1	5	27	127	78
Organic matter (g/dm⁻³)	194	153	143	120	97
C/N ratio	30	26	19	22	15
Number of plant species in community	12	20	19	24	30

d: dry soil
C/N ratio: ratio of weight of carbon to that of nitrogen
Source: Miles, 1981

3. (a) Describe the vegetation patterns shown in Figure 3.1.
   (b) Using a good atlas, construct distribution maps of relief and altitude (i.e. land over 250 metres) and average rainfall for Great Britain.
   (c) Explain the distribution of vegetation/land-use in relation to altitude, relief and rainfall conditions.
   (d) Refer to Figure 2.21 on page 58. What is the link between vegetation and general soil type?
   (e) How does human activity affect the relationship between vegetation and physical factors in the uplands?

4. Project themes

   A list of possible themes for project work is given here. Details of methodology or approach depend on the specific location and the time and resources available, so they must be worked out between teacher and students.

   (a) Comparative analysis of vegetation of (i) heather moorland; (ii) grassy heaths; (iii) blanket bog.
       Relation of these plant communities to soil, e.g. (i) podzols; (ii) peaty podzols with gleying; (iii) peats.
   (b) Transect analysis of vegetation change along slopes. Relationship of species found with: (i) slope; (ii) altitude; (iii) drainage; (iv) pH; (v) amount of soil organic matter; (vi) amount of soil moisture (see Chapter 8).
   (c) Plant succession
       (i) Vegetation analysis of dune system; fresh water ponds.
       (ii) Colonisation of heather moorland by woodland (pine or birch); rough pastures by bracken.
       (iii) Vegetation colonisation of waste ground; refuse tips.
   (d) Human impact
       Investigation of effects of trampling on vegetation across footpaths, affecting: (i) numbers of species; (ii) growth form of plants; (iii) height of vegetation; (iv) percentage cover of vegetation; (v) soil compaction and erosion.
       This exercise can be linked with Project 2 on pages 181, 182.
   (e) Litter exercise
       Comparison of coniferous (e.g. pine) and deciduous (e.g. oak, sycamore) woodland litter. Collection of 200 leaves from each woodland. Measurement for each litter of: (i) moisture content; (ii) ash content; (iii) rate of decomposition. Analyses of soil beneath coniferous and deciduous woodland. Relation of litter characteristics to observed soils.

4 Ecosystems

Up till now we have examined soils and vegetation separately. We have looked at them more or less as isolated units within the larger environment. When we examine the total environment, including soils, vegetation, animals, humans and climate, together with their interactions, we study what are called *ecosystems*.

Ecosystems consist of living organisms (the biotic community or component) and their physical and chemical environment (the inorganic component). They are identified by their structure and function. By *structure* we mean the arrangement of their physical parts which we can see, feel and experience with all our senses. Thus, all species of plants and animals, together with their waste products and dead remains in the soil, are part of an ecosystem structure. Also included are the non-living rocks and sediments which lie beneath ecosystems and the climatic factors (light, heat, moisture) which act upon them. By *function* we mean all the processes and interactions which take place within the ecosystem. As we shall see, such processes help to unite the diverse component parts of the ecosystem into a single or integrated unit.

It is convenient to distinguish ecosystems on the basis of their components that can been seen. This is most often done by using vegetation. Thus, ecosystems can be viewed as distinct vegetation units, such as an oak wood, a tropical rain forest, temperate grassland or desert scrubland. Together with their animal populations (including humans) and habitat conditions (soil, terrain, *microclimate*), they constitute well-defined ecological units. (See Figure 2.3 on page 32.)

A. Ecosystem Structure

Chapter 5 discribes the principal structural components of the major ecosystems of the world, giving an outline of their unique plants, animals, soils and climates. In this section, we shall limit ourselves to an examination of the two organic components common to all ecosystems: biomass and dead organic matter.

1. Biomass = living weight

The term *biomass* refers to the total amount of living matter present at any given moment in an ecosystem. It is usually expressed as a dry weight of tissue per unit area, e.g. tonnes/hectare or kg/m^2. The living organisms of a particular ecosystem – the plants, animals and soil micro-organisms – show a characteristic biomass or *standing crop*. For instance, the vegetation biomass of the tropical rain forest, dominated by a large system of trees, is very much greater than that of the temperate grasslands of North America, or of the tundra vegetation of Arctic regions which is composed of mosses, lichens, hardy grasses and low woody shrubs (see Table 4.1). Notice that biomass is present below ground (roots, animals, micro-organisms) as well as above the surface (stems, leaves, animals). In tropical and temperate forests, three-to-four times as much vegetation biomass occurs above the ground as below it. In the tundra and temperate grasslands, the above-ground/below-ground biomass ratio is different. There is between four-to-nine times respec-

Table 4.1 Biomass and productivity in four selected ecosystems

Ecosystem	Biomass (kg/m²)	Above-ground to below-ground biomass ratio	Average productivity (kg/m²/yr)
Tropical rain forest (Amazonia, Brazil)	45	4:1	2.2
Temperate deciduous forest (oak woodland, UK)	30	3:1	1.2
Temperate grassland (prairies, North America)	1.6	1:5–1:9	0.6
Arctic tundra (Alaska)	0.6	1:4–1:5	0.14

Figure 4.1 Plan of an ecosystem, showing the distribution of the principal structural components, e.g. plants, animals, micro-organisms, soil and climate. Functional exchanges of materials and energy between the ecosystem and its surrounding environment are also indicated. (Source: Walter, 1973)

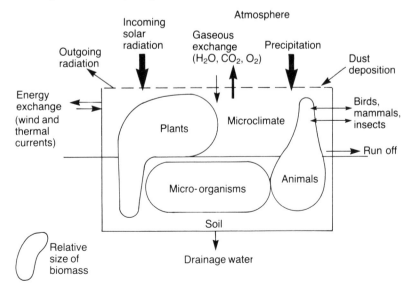

tively as much living vegetation biomass below ground (roots, horizontal stems) as above (stems, leaves).

Figure 4.1 shows the form and function of an idealised ecosystem. Although most of the plant biomass is above the ground (as in the tropical rain forest), the reverse is true for animals and soil micro-organisms. Despite the fact that we see so much animal life above the surface, most animal biomass (earthworms, insects) is found below ground-level, in the soil.

2. Dead organic matter

Dead and decaying plant and animal remains are an important component of individual ecosystems (see Chapter 1, Section C). In many ecosystems dead organic matter (DOM), made up of surface litter and soil humus, greatly exceeds the living biomass in volume or weight. This point is well illustrated in Figure 4.2 for a range of ecosystems. While the quantity of DOM (soil humus and litter) just exceeds that of the living biomass (leaves, branches, stems, roots) in the temperate deciduous

Figure 4.2   The distribution of dead organic matter (litter and humus) and living biomass (surface stems/leaves and roots) in three selected ecosystems.

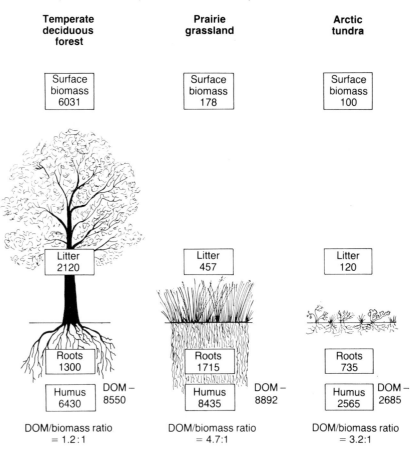

forest, the former is the major component in the tundra and temperate grassland ecosystems. In the tundra, DOM exceeds living biomass by over 3:1 and in the temperate grassland the ratio increases to almost 5:1.

ASSIGNMENTS
1. (*a*) *Define what is meant by an ecosystem.*
    (*b*) *Distinguish between the structure and function of an ecosystem.*
2. *Refer to Table 4.1 and Figures 4.1 and 4.2.*
    (*a*) *What is: (i) biomass; (ii) dead organic matter?*
    (*b*) *How can they vary: (i) within ecosystems; (ii) between ecosystems?*

B. Ecosystem Processes (Function)

All ecosystems have two major processes: a *flow* (of energy) and a *cycle* (of nutrients). These two processes, which are driven by the sun's energy, help to link the component parts of ecosystems.

1. Energy flow

Individual ecosystems, whether we are dealing with the tropical rain forest or a small pond, are sustained by a flow of energy through them. The main source of this energy is sunlight or solar radiation. Light energy enters the ecosystem when it is absorbed by green plants (see Figure 4.3). It is then passed through the ecosystem as food; plants are eaten by animals and animals consume each other. Energy absorbed and transferred through the ecosystem in this way is eventually converted into heat. This heat, which is exactly equivalent to the solar input, finally leaves the ecosystem and is lost in space.

(*a*) *Primary production*

Green plants are able as they grow to absorb light energy from the sun and raw materials from the environment (nutrients in the soil solution, gases in the atmosphere). The process of absorbing light energy (or light *fixation*) by plants is called *photosynthesis* or *primary production*. It has the formula:

Figure 4.3 Simple model showing how an ecosystem is sustained by inputs and outputs of energy at its boundaries

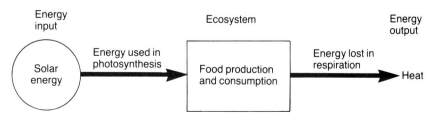

97

Figure 4.4 Fate of solar energy fixed by green plants in an ecosystem

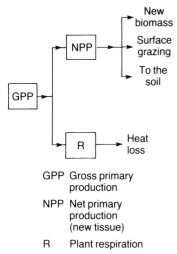

GPP Gross primary production

NPP Net primary production (new tissue)

R   Plant respiration

$$6CO_2 + 6H_2O + \text{light energy} = C_6H_{12}O_6 + 6O_2$$

| carbon dioxide | water | | sugar (carbohydrate) | oxygen |

A distinction is made between gross primary production (GPP) and net primary production (NPP). The former is the total amount of energy absorbed or fixed by green plants (see Figure 4.4). Some of the GPP goes into the production of new plant tissue. This is the NPP. The remaining part of the GPP is re-used by plants to carry out normal functions, e.g. absorbing nutrients, repairing cell damage. This re-used energy, which is made available by a process known as *respiration*, is eventually lost as heat. As shown in the formula below, respiration is the converse (opposite) of photosynthesis:

$$C_6H_{12}O_6 + 6O_2 = 6CO_2 + 6H_2O + \text{release of energy}$$

| sugar (new production) | | oxygen | carbon dioxide | water | respiratory heat (lost to ecosystem) |

The rate of energy fixation or production is referred to as *productivity*. It is expressed in units as 'energy stored per unit of ground area per unit of time', e.g. tonnes/hectare/year or $kg/m^2/year$. As a *rate* of production of new tissue or biomass, productivity should not be confused with total biomass. As suggested on page 95, the standing crop or biomass is the total amount of living tissue accumulated over a period of time, expressed at any one moment. The huge biomass of the tropical rain forest is not equivalent to the annual productivity, which happens to be high (Table 4.1), because that biomass probably took hundreds of years to accumulate.

Plant productivity is determined by light, temperature, water, nutrients and carbon dioxide. It is highest where light, warmth, moisture and key nutrients are abundant, e.g. in swamps, marshes and, as shown in Table 4.1, tropical rain forest. It is lowest in areas where light, warmth and/or water are lacking, e.g. in the Arctic tundra and the hot sandy deserts. There is a more detailed examination of the variation in plant productivity among the major terrestrial ecosystems of the world in Chapter 6, Section A.

(b)   Trophic levels

Plant growth, represented by NPP, is the principal food resource for most animals and soil organisms in the ecosystem. As shown in Figure 4.4, a proportion of the NPP may be stored for a time as new biomass (e.g. in the trunks and branches of trees), while some may be consumed directly by animals above the surface of the ground. On land, the majority of NPP, however, enters the soil where it will form humus and will eventually be eaten by soil organisms. Ultimately, even the long-lasting trees will die and be consumed by animals. In summary, therefore, plant material above and below the ground surface is eaten by herbivores (plant eaters), which may then be consumed by carnivores (meat eaters). Plants and animals are also eaten by soil micro-organisms, especially by bacteria and fungi which are referred to as the *decomposer organisms*.

The movement or flow of energy from plants to animals to decomposers can be represented in its simplest form as a *food chain* (see Figure 4.5). Each stage in the chain where energy is exchanged is called a *trophic level* or 'feeding' level. Thus, plants (e.g. alfalfa grass) are represented by trophic level 1, herbivores or primary consumers (e.g. beef cattle) by trophic level 2, and carnivores or secondary consumers (e.g. humans) by trophic level 3. The decomposer part of the system can be in trophic levels 2, 3 and 4 depending on the source of the food. Decomposers receive dead material from trophic levels 1, 2 and 3 and waste animal products from levels 2 and 3.

There may be one or two additional levels in a food chain represented by 'top' carnivores. These are carnivores which eat other carnivores, e.g. in other food chains a fox may eat a snake, magpies may eat small cats. Few food chains, however, involve more than five or six trophic levels because a significant amount of energy (as much as 90%) is lost as heat at each step in the food chain. There is seldom enough energy available, therefore, to support an additional level (see Figure 4.6).

Although simple linear food chains of the type shown in Figure 4.5 seldom exist in nature, something close to them occurs in species-poor Arctic ecosystems; as in the sequence of algae in the ocean through an algae-eating fish to a sea lion to a polar bear. In reality, the situation is often more complex because many organisms including humans operate at several trophic levels. For instance, a person might eat the bear, the seal, the fish or even the algae, placing him in trophic levels 5, 4, 3 or 2. Or the polar bear might eat the fish or the person. For these reasons it is more realistic to describe energy flow in terms of a *food web* rather than as a food chain (see Figure 4.7).

Figure 4.5 Flow of energy through an ecosystem in the form of a food chain, i.e. plant → herbivore → carnivore. The point where food energy is exchanged is called a feeding or trophic level.

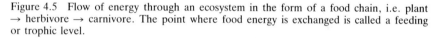

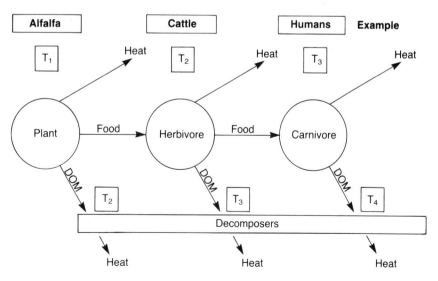

T Trophic level   DOM Dead organic matter

Figure 4.6 Effect of heat loss on trophic level structure. Because of considerable heat loss at each trophic level, the numbers of individuals, the biomass and the productivity of a trophic level are always much less than the preceding one and much larger than the next. As a result, the trophic distribution of energy in the ecosystem is pyramid-shaped. This is shown by using the hypothetical food chain of alfalfa → cattle → young human. (Source: Southwick, 1976)

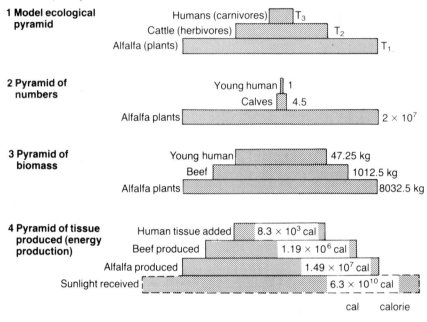

(c) Humans and energy flow

Humans are able to modify greatly the flow of energy in an ecosystem (see Figure 4.5). In many human-regulated ecosystems, energy flow is altered so that *yields* are as high as possible. Yield refers to the rate at which an ecosystem produces useful products. One objective of management is to increase yield by channelling as much net production to the product in question as is possible. Thus, a forester may be interested in wood yield, a farmer in the yield of potatoes or grain or beef (see Figure 4.5), a fisherman in fish yield.

The yield of certain desired products can be enhanced by: (i) ensuring good growth conditions, e.g. adding fertiliser and water, using good seed; (ii) preventing, as far as possible, loss of yield by consumption and competition, e.g. reduction of grain yield by disease attack, consumption by birds and small mammals, or weed invasion.

2. Nutrient cycling

(a) Stores and pathways

Biological or *nutrient cycling* refers to the circulation of chemical elements from environment to organisms and back again to the environment. Figure 4.8 illustrates a simple model of nutrient circulation. Nutrient elements are stored in three compartments within the ecosystem: in the soil, in the living biomass and in surface litter. Nutrients

Figure 4.7   Simple food web in an Arctic terrestrial ecosystem

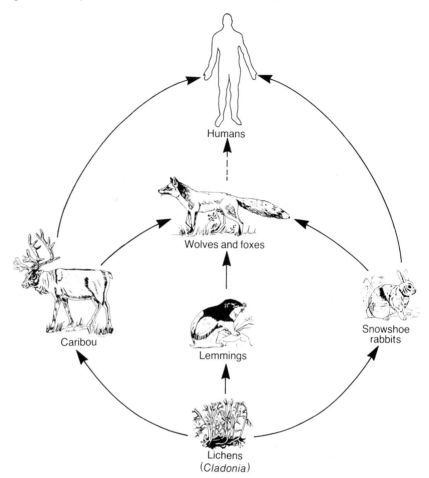

Humans

Wolves and foxes

Caribou

Lemmings

Snowshoe rabbits

Lichens
(*Cladonia*)

Figure 4.8 Model of nutrient circulation, input and loss in an ecosystem (Source: Gersmehl, 1976)

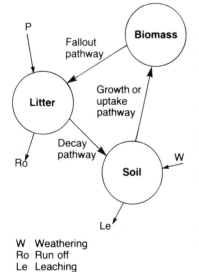

W Weathering
Ro Run off
Le Leaching
P   Precipitation

are recycled between environment and organisms via three main pathways. (1) In the 'uptake' or 'growth' pathway, simple inorganic elements and compounds of nitrogen, phosphorus and potassium, together with those of other essential elements, are taken up from the soil and converted into complex, organic substances within the living biomass. During this growth phase other elements, e.g. carbon and oxygen from the atmosphere and hydrogen in water, are also incorporated. (2) As plants and animals die, they contribute nutrients to the litter store. This is the 'fallout' pathway. (3) A 'decay' pathway is formed by the decomposition of litter to humus, with the eventual release of inorganic nutrients back into the soil.

The storage of nutrients within an ecosystem is not a static condition. Nutrients can increase or decrease, often rapidly, over a period of time. For this reason the three-compartment/pathway model in Figure 4.8 is not self-contained or 'closed'. Other pathways exist, which add nutrients to or remove nutrients from the system. For instance, chemical elements are increased within the soil compartment by rock-weathering. They may also

101

enter the ecosystem from other ecosystems in rainfall (precipitation) and as wind-blown dust and leaves. Animals moving from one ecosystem to another may also alter the respective balance of nutrients. On the other hand, chemical elements can be lost from one ecosystem to adjoining ecosystems mainly by run off (erosion) and the leaching of soil.

(b) Humans and nutrient cycling

Humans can alter fairly easily the composition, storage and circulation of chemical elements within ecosystems. An increase in the storage of nutrients occurs when inorganic fertilisers, rich in nitrogen, phosphorus and potassium, are added to soils. Provided that the amounts are not excessive, such fertilisers will increase crop yields and thus speed up the rate of nutrient cycling. A good example of an indirect exchange of chemical elements from one ecosystem to another is sulphur from coal-burning power stations in Britain being carried by winds to Scandinavia. Many habitats there, including lakes and soils, are being made more and more acid by excessive sulphur deposition, mostly in the form of acid rain or weak sulphuric acid (H_2SO_4). This has resulted in a decrease in fish and a decrease in the growth of crops and forests, and thus in a decline in the rate of nutrient cycling. (See Chapter 7, Section B.)

Human activity may reduce the nutrient resources of an ecosystem by misusing and over-exploiting the land. Figure 4.9 illustrates one such example where nutrients are extracted at a rate faster than they are replaced. In the Third World, two-thirds of the population relies on wood for cooking and heating, but almost half of these people face shortages of fuel wood. To meet their needs, they are forced to burn firstly the crop stalks, which remain after the harvest, and then, when the crop stalks have run out, animal dung. Normally the crop stalks would be fed to the animals, whose dung could be used as a fertiliser for the next crop. This would recycle some of the nutrients taken out of the soil in crop (biomass) harvest. But the shortage of fuel wood means that the fertility of the topsoil is not being replaced. Once deprived of the nutrients and the organic remains needed for strutural stability, topsoil erodes more easily.

ASSIGNMENTS
1. (a) *Define what is meant by energy flow.*
    (b) *Using Figure 4.4, distinguish between gross and net primary production.*
    (c) *What is plant productivity? How does it differ from biomass (see Table 4.1).*
    (d) *Give five factors which affect plant productivity.*
2. (a) *Using Figures 4.5 and 4.6, explain the terms: food chain, trophic level.*
    (b) *Arrange the following organisms on a labelled diagram of a food chain and explain your choice of sequence: field mouse, soil bacteria, grass, eagle owl, weasel.*

Figure 4.9 Nutrient cycling in Third World farming systems showing: (a) balanced nutrient cycling with nutrients extracted at the same rate as they are replaced; (b) unbalanced nutrient cycling due to the removal of forests and the shortage of fuel wood (Source: Hartzell, 1986)

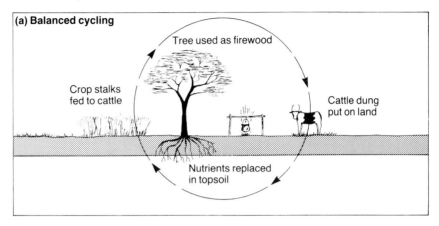

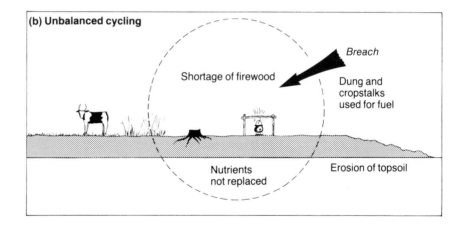

    (c)   *Why are food chains seldom longer than four or five trophic levels?*

3.   *Refer to Figures 4.7 and 4.10.*
    (a)   *What do these diagrams tell us about the pathway of energy through the ecosystem?*
    (b)   *Give an example from Figure 4.10 of: (i) a relationship between producer, primary consumer and secondary consumer; (ii) a relationship with five trophic levels.*

4.   (a)   *Define nutrient cycling.*
    (b)   *Using Figure 4.8, describe how nutrients are: (i) stored; (ii) recycled within ecosystems.*
    (c)   *Refer to Figure 4.9. Give three ways that nutrients can be: (i) increased; (ii) lost from ecosystems.*

103

Figure 4.10   Model of a food web in a temperate hardwood forest, north east USA

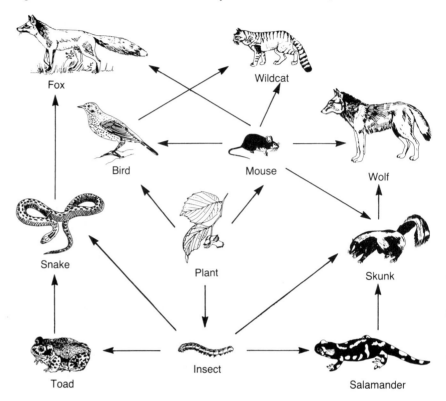

Key Ideas

Introduction

1. An ecosystem consists of a set of living organisms and their physical and chemical environment.
2. By the structure of an ecosystem we mean the arrangement of its physical components, e.g. plants, animals, soils, climate.
3. By the function of an ecosystem we mean how it works, how its various components are linked together into a single integrated unit.
4. Ecosystems can be identified by using vegetation, e.g. tropical rain forest, Alpine pasture, salt marsh.

A.   Ecosystem Structure

1. Biomass is the total amount of living material in an ecosystem at any given moment.
2. It can be measured as a (dry) weight of tissue per unit area.
3. Biomass is distributed unevenly: (i) within individual ecosystems, as indicated by the above- to below-ground biomass ratio; (ii) between different ecosystems.

4. In many ecosystems the dead organic matter (DOM) greatly exceeds the living biomass in volume or weight.

B. Ecosystem Processes (Function)

1. Energy flow involves: (i) the absorption of light energy by green plants; (ii) the passage of this energy through the ecosystem as food by means of animal consumption; (iii) its final loss from the ecosystem as heat.
2. Energy fixation or light fixation by plants is called photosynthesis or primary production.
3. The rate of energy fixation or production is referred to as productivity.
4. The simple linear flow of energy in the ecosystem from plants to animals to decomposers is called a food chain.
5. Each stage in the chain where energy or food is exchanged is known as a trophic or feeding level.
6. More realistically, energy flows through an ecosystem in the form of a more complicated food web.
7. There are seldom more than five or six trophic levels in a food chain or web because respiratory heat is lost at each trophic exchange.
8. Humans modify energy flow within ecosystems by making the production or the yield of desirable products as high as possible.
9. Biological or nutrient cycling refers to the circulation of chemical elements from the environment to organisms and back again to the environment.
10. The reserve of nutrients in ecosystems can be easily increased or decreased by human activity.

Additional Activities

1. Refer to Table 4.2.
    (a) Describe the distribution of the total storage of organic matter, indicating where the greatest store lies.
    (b) Using the text, suggest whether this distribution is typical or not in the ecosystem concerned.
    (c) Using the data in Section B, construct an energy flow diagram similar to that in Figure 4.4.
    (d) Calculate the percentage of solar radiation used in: (i) photosynthesis; (ii) transpiration (movement of water from soil to leaves to atmosphere). Comment on your results.
    (e) What percentage of the total energy fixed annually by the forest goes into the production of new tissue?
    (f) Describe the fate of this new production.
    (g) What evidence suggests that this forest is still growing?
2. (a) Examine the model in Figure 4.11.
    (b) Describe the condition of nutrient storage and circulation indicated by each of the four boxes.

(c) Illustrate each condition using one of the following examples: over-cropping, addition of nitrogen fertiliser, deposition of acid rain, burning heather moorland.

(d) Give reasons for your answers.

Table 4.2 The storage of organic matter and the annual energy flow in a deciduous woodland in New Hampshire, USA

		Ecosystem attribute	kCal/m²
Section A Total storage	1	Living biomass (a) above ground (b) below ground	71.4 59.7 11.7
	2	Litter	34.3
	3	Soil DOM	88.1
	4	Annual solar radiation	1010.0
	5	Energy used in: (a) Photosynthesis (b) Transpiration	10.4 175.6
Section B Annual energy flow (during one year)	6	Gross primary production (a) Respiration (b) Net primary production	10.4 5.7 4.7
	7	Annual net primary production consumed by animals	3.5
	8	Annual biomass storage (a) above ground (b) below ground	1.2 0.95 0.25
	9	DOM storage	0.15

Source: Gosz *et al*, 1978

Figure 4.11 Four-compartment model of human impact on the nutrient cycle (Source: Tivy and O'Hare, 1981)

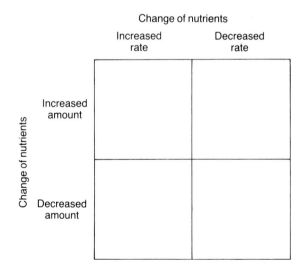

5 The Global Ecosystems: Structure and Distribution

A. Introduction

The structure and distribution of the world's principal ecosystems are the subject of this chapter. The major global ecosystems, such as tropical rain forest, temperate grassland, tropical desert, are described in terms of their associated climates, soils, vegetation and animal life. The ways in which human activity can alter the structural features of ecosystems are examined and illustrated by the clearance of the Mediterranean forest and of the tropical rain forest.

Figure 5.1 World distribution of the main zonal soils

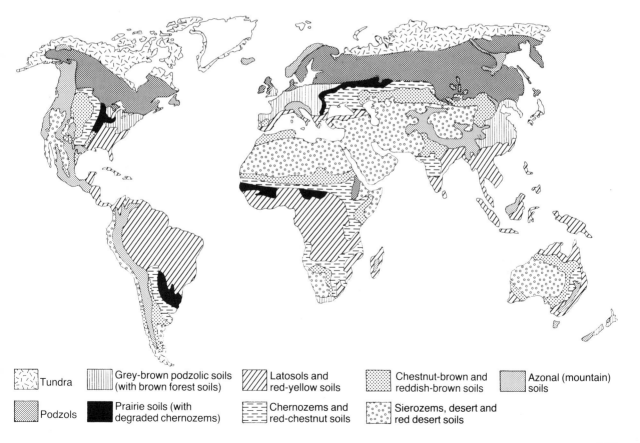

Tundra	
Podzols	
Grey-brown podzolic soils (with brown forest soils)	
Prairie soils (with degraded chernozems)	
Latosols and red-yellow soils	
Chernozems and red-chestnut soils	
Chestnut-brown and reddish-brown soils	
Sierozems, desert and red desert soils	
Azonal (mountain) soils	

1. Climate, soils and vegetation: zonal arrangement

As mentioned in Chapter 2, Section C.4, many mature, 'steady-state' soils are strongly influenced by climatic conditions. At the global level there is a series of major soil groups, called *zonal soils* (see Figure 5.1), which are determined by a limited range of present-day climatic types or zones (see Figure 5.2). There is also a close link between the major climatic zones and the distribution of climax vegetation formations (see Figure 5.3).

At the global level, therefore, there is a close zonal relationship between climate and soils and vegetation. This relationship is summarised in Figure 5.4.

2. The biome framework

The world's zonal relationships between climate and soils and vegetation can be studied by using the idea of the ecosystem. The most extensive world-scale ecosystem which can usefully be named is the *biome*. Biomes are large-scale ecosystems defined or identified by their dominant vegetation cover (e.g. deciduous forest, grassland, desert). Thus, a map of the major world biomes is really just a vegetation map (see Figure 5.3). By defining the outlines of ecosystems in this way, biomes can be used to show the relationships between climate and soils and vegetation.

Figure 5.2   Main thermal zones and bioclimatic regions of the world

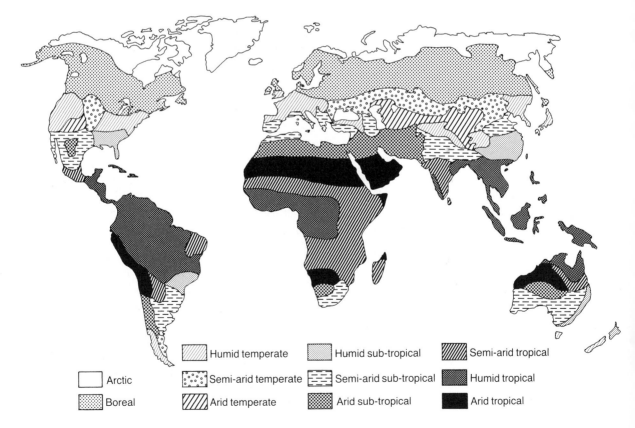

Arctic

Boreal

Humid temperate

Semi-arid temperate

Arid temperate

Humid sub-tropical

Semi-arid sub-tropical

Arid sub-tropical

Semi-arid tropical

Humid tropical

Arid tropical

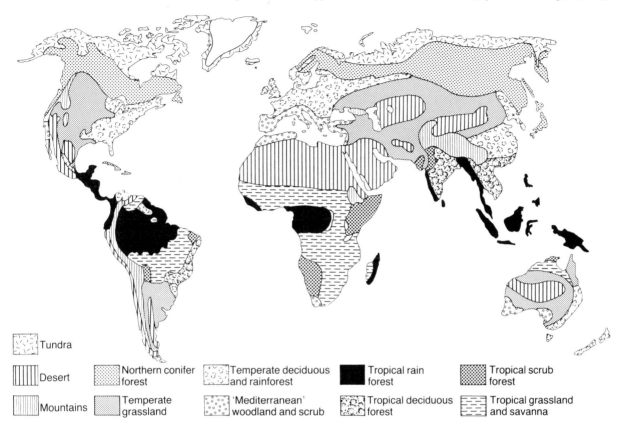

Figure 5.3 Major biomes of the world. Many have been greatly modified by human activity (e.g. temperate grassland, temperate deciduous forest, tropical rain forest). Some may owe their origin and overall appearance to human intervention (e.g. the savanna grasslands).

Tundra

Desert

Mountains

Northern conifer forest

Temperate grassland

Temperate deciduous and rainforest

'Mediterranean' woodland and scrub

Tropical rain forest

Tropical deciduous forest

Tropical scrub forest

Tropical grassland and savanna

3. Climatic diagrams

In order to relate the various climates of the major biomes to their respective soils and vegetation, we can use a framework showing climatic features (see Figure 5.5). Climatic patterns are represented by using average monthly temperatures and amounts of rainfall. Where monthly rainfall (1 unit of the vertical axis = 20 mm) 'exceeds' monthly temperature (1 unit on the vertical axis = 10 °C), a relatively humid season results. On the other hand, a relative drought occurs when the reverse applies, with temperatures 'greater' than rainfall. This model seems very simple but it has been shown to be fairly accurate. The value of potential evapotranspiration data in the calculation of wet and dry periods (in a ratio with rainfall for instance) is not used here because the calculation of evapotranspiration in many areas outside North America and Europe (e.g. the tropics) is either unavailable or unreliable.

In Figure 5.5 there is a change of scale for those months with over 100 mm of rainfall. Monthly rainfall values in excess of 100 mm are represented on a scale of 1 unit = 200 mm (i.e. ten times the standard rate) and are shown in black.

Figure 5.4 Relationships at the global level between climate, soils and vegetation, showing main interactions

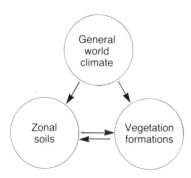

General world climate

Zonal soils

Vegetation formations

109

Figure 5.5 Key to all of the climatic diagrams used in this text.
Horizontal axis: months
Vertical axis: one division = 20 mm of rainfall or 10 °C
a: weather station
b: height above sea level (metres)
c: mean annual temperature in °C
d: mean annual precipitation in mm
e: curve of mean monthly temperature
f: curve of mean monthly precipitation
g: relative period of drought
h: relative humid season
i: mean monthly rain with over 100 mm (scale: one division = 200 mm
j: months with mean daily minimum below 0 °C (cold season)
k: months where late or early frosts occur (absolute minimum below 0 °C)
(Source: Walter, 1973)

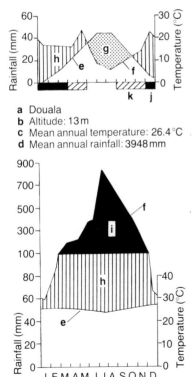

a Ankara
b Altitude: 895 m
c Mean annual temperature: 11.7 °C
d Mean annual rainfall: 341 mm

a Douala
b Altitude: 13 m
c Mean annual temperature: 26.4 °C
d Mean annual rainfall: 3948 mm

ASSIGNMENTS

1. (a) What is a biome?
   (b) Using Figures 5.1–5.3, describe the world distribution of: (i) soils; (ii) vegetation; (iii) climate.
   (c) Using Figure 5.4, examine the functional relationship between the three elements.
2. Refer to Figure 5.5.
   (a) Describe the way in which temperature and rainfall are recorded in the given model.
   (b) For each climatic station calculate: (i) the duration of the humid season; (ii) the duration of the dry season; (iii) the number of months with mean temperatures below 0 °C; (iv) the average length of the frost-free period.
   (c) Locate each of the given stations on a world map and comment on your results.

B. Temperate Forest Biomes

Three main forest biomes are located in the temperate zone: (i) boreal (northern) coniferous forest of cold temperate latitudes; (ii) cool temperate, deciduous forest, which formerly covered much of Europe, north China and eastern North America; (iii) warm temperate, mixed forests of conifers and broad-leaf evergreens (e.g. evergreen oak), which once covered much of the Mediterranean lands but of which little now remains (see Figure 5.3).

1. Boreal forest biome

(a) Distribution

The northernmost temperate forest biome is the boreal forest or *taiga*. This is one of the largest plant formations on earth and extends in an unbroken belt across the whole of northern North America and Eurasia, south of the tundras (see Plate 5.1). Taiga is also found on high mountains in lower latitudes, such as the southern Rockies. There is nothing like it in the southern hemisphere because there are no large land-masses occupying similar latitudes to those of the northern hemisphere where the forest is found.

(b) Climate

In the northern forests the climate is *cold continental*, with cool summers, prolonged harsh winters and a variable but usually adequate total precipitation, most of which occurs in the latter part of the summer (see Figure 5.6). Winter is the dominant season. Many areas have at least six months with average monthly temperatures below freezing-point. The frost-free period in the taiga is therefore short. In interior continental stations (e.g. northern Siberia) it lasts for only 50–100 days but rises to over 250 days

110

Figure 5.6 Characteristic climate of the boreal forest biome, showing a long severe winter, a short summer and a high range of temperature (Source: Walter, 1973)

Archangel
Altitude: 10 m
Mean annual temperature: 0.4 °C
Mean annual rainfall: 466 mm

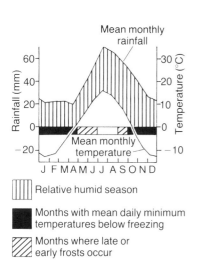

||||| Relative humid season

■ Months with mean daily minimum temperatures below freezing

▨ Months where late or early frosts occur

Plate 5.1 Boreal forest biome, Finland. This forest is composed mainly of evergreen conifers. The shrub layer is typically little developed with some young tree seedlings, low woody shrubs and a carpet of pine needles. (Photograph: Barnaby's Picture Library/ S. Nickels)

in areas close to the ocean. To some extent, the short growing season of summer is offset by quite warm and very long summer days.

Because of the low temperatures of this biome, much of the region is affected, at or near the surface, by permanently frozen ground called *permafrost*. However, snowfall is heavy and snow-cover long-lasting. A thick snow-cover insulates the ground from too much heat loss, and thus helps to keep surface soil temperatures just below freezing.

(c) Vegetation response

Trees are the dominant plant form in this zone of cold continental and sub-Arctic climate. They are mostly evergreen conifers with needle-shaped, wax-covered leaves. Such trees are able to: (i) photosynthesise all year if temperatures are high enough; (ii) resist drought from strong winds and frozen soil; (iii) remain undamaged by snowfall because of their overall conical shape.

Variety of species in this biome is low. Although higher than that in the tundra to the north, it is much lower than that in the temperate deciduous forest biome, for instance. The dominance of trees is maintained over large areas by four types of conifer tree, i.e. spruce, pine, fir and larch, with some examples of tolerant deciduous hardwoods, e.g. alder, birch and poplar. There are important regional variations in climate, soil and vegetation within the biome. In arid mountains, for instance, open pine woodland occurs in the sub-Alpine zone; in lowland, humid sub-Arctic areas (e.g. the lower reaches of the River Ob in northern USSR), extensive bogs occur; in very cold, dry areas (e.g. northern Siberia), great stands of deciduous (leaf-shedding) larch are found. (Larch is a deciduous conifer!)

(d) Animals

Animals in this biome are limited by severe winters and the small number of different habitats available. The stratification of the forest, for instance, is fairly simple. Apart from the tree canopy, the understorey consists of a number of low, berry-bearing shrubs (e.g. crowberry, blueberry), mosses, lichens or a thick carpet of pine needles (see Plate 5.1). The most important herbivores are deer. The moose is typical of the North American coniferous forest. Rodents are also plentiful, and can avoid the harshness of the winter by burrowing under the snow. Carnivores include wolves, lynxes, weasels, mink and sable. Birds are relatively few and are either adapted to feeding in the taiga (e.g. the crossbill) or are summer migrants, feeding on the vast, seasonal swarms of insects.

(e) Soils

Most of the soils of this biome belong to the podzolic soil group (see Chapter 2, Section B.3).

2. Temperate deciduous forest biome

(a) Distribution and climate

Temperate deciduous forests are found mainly in north-west, central and eastern Europe, eastern North America and east Asia (e.g. northern China, Korea, Japan).

Compared with the climate of the boreal forest, that of the deciduous forest zone is much more favourable for plant growth and development. A more southerly location, close to the oceans (see Figure 5.7), results in higher overall temperatures and a much longer growing season lasting 4–7 months (as opposed to 3–4 months in the boreal biome). Winters are not as cold as in the boreal zone, although average temperatures often fall below zero except in regions close to the sea. This biome is well supplied with rainfall, which is well distributed throughout the year.

(b) Vegetation response

The main feature of the trees of this biome is that they are deciduous, i.e. they shed their leaves in the winter months as a response to reductions in light and heat. They shed their leaves to conserve moisture. The leaves of the trees are typically broad-leaf. Being neither thick, heavy nor waxy, they have no special protection against moisture loss by transpiration (cf the boreal forest), which would not be made up from a cold soil during the winter. The shedding of leaves checks this.

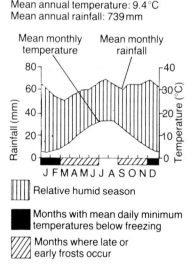

Figure 5.7 Climatic conditions typical of the temperate deciduous forest biome, with well-distributed precipitation, a cold winter and a warm summer (Source: Walter, 1973)

Luxembourg
Altitude: 362 m
Mean annual temperature: 9.4 °C
Mean annual rainfall: 739 mm

JFMAMJJASOND

Relative humid season

Months with mean daily minimum temperatures below freezing

Months where late or early frosts occur

Broad-leaf trees of oak, ash, elm, beech, maple and others are dominant in the deciduous forest biome (see Plate 5.2). The system is much more varied than in the boreal biome. Numbers of tree species per hectare can be between five and twenty, and the forest commonly has a four-tier vertical stratification, as shown in Figure 3.3.

(c) Animals

Animal life – mammals, birds and insects – is also more varied than in the boreal biome. In the USA, white-tailed deer and black bear are characteristic large mammals; foxes, polecats, squirrels, badgers are typical smaller animals. This biome is extremely rich in bird species, especially owls, woodpeckers, thrushes, warblers and finches.

(d) Soils

The characteristic soils of the temperate deciduous forests are brown soils and similar soils (see Chapter 2, Section B.2).

3.   Mediterranean forest and scrubland

(a) Distribution and climate

The major area of this biome is the Mediterranean basin but other

Figure 5.8 Mediterranean type of climate, as at Messina, southern Italy, which shows the hot, dry summer conditions and the moderate rainfall of winter, spring and autumn (Source: Walter, 1973)

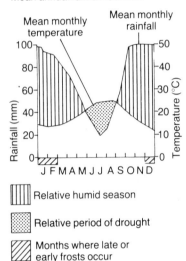

Messina
Altitude: 60 m
Mean annual temperature: 18.3 °C
Mean annual rainfall: 862 mm

|||| Relative humid season

▓▓▓ Relative period of drought

▨▨▨ Months where late or
      early frosts occur

regions where it occurs include California, south-west South Africa and south-west Australia (see Figure 5.3).

The climatic conditions typical of this biome are shown by Messina, Italy (see Figure 5.8). There is a marked seasonal swing between mild, wet winters and hot, dry summers. Winter and spring rains produce an accumulation of moisture in the soil which plants use during their most active growth period in the spring, when temperatures are rising. Shortages of soil water (or droughts), which develop during the long, dry summer, put a check on plant growth and result in special adaptations by plants.

(b) Vegetation response

Originally the dominant vegetation was mixed forest of conifers (e.g. Aleppo pine) and broad-leaf evergreens (e.g. evergreen oak). Because of human interference little of this forest now remains. The characteristic vegetation is now a mixture of tall shrubland, low scrub and grassland.

Most plant species of this biome, whether trees or shrubs, have a range of adaptations to resist or evade drought. (1) Many are evergreen, with hard (i.e. sclerophyllous), thick, waxy leaves to reduce water loss by transpiration when water is scarce (e.g. *Arbutus* tree, Mediterranean heath). (2) Some plants, such as the Mastic tree, are able to close their stomata (i.e. the leaf vents from which water is transpired) in the summer drought period. (3) The chamise is a plant which reduces transpiration loss by having very small leaves. (4) Some reduce their leaf surface in the form of thorns or spines, as with the succulents (see page 128) of the cactus family. (5) Several plants adapt to drought by tapping underground water supplies through deep root-systems (e.g. Almond tree). (6) Others, in the form of bulbs and tubers, remain dormant beneath the soil during the dry summer, but burst into growth in the wet spring to flower (e.g. hyacinth, crocus). (7) A number of plants, including the annual grasses, survive the summer drought in the soil as seeds and spores and complete their growth cycle during the cooler, more moist period.

(c) Human impact on vegetation

As a result of long and fairly intensive usage by humans (e.g. forest clearances, cultivation, grazing), many Mediterranean environments are now very degraded. They show a classic sequence of deterioration in vegetation and soil (see Figure 5.9).

(i) *Maquis.* The original forest which cloaked the lands surrounding the Mediterranean basin sheltered a tall shrub-layer (3–5 metres in height), including such plants as box, viburnum and species of rose. Beneath there was also a herb-layer (50 cm), made up of species of *Rubia*, *Asparagus* and *Carex*. Some forest survives today but, in general, Mediterranean landscapes are made up of shrubby vegetation, cultivated trees (olives, vines, almond) and local areas of intensive farming. Many of the shrubby

Figure 5.9 Sequence of deterioration in vegetation and soil in 'Mediterranean' environments subject to increasing levels of human activity

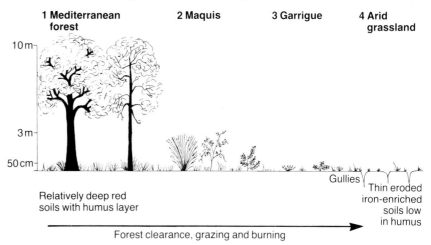

plants now form dense thickets as a result of the destruction of the original forest by human activity, grazing animals or fire. The taller shrub communities bear local names, e.g. the maquis of Mediterranean Europe (see Plate 5.3 on page 76), the chaparral of California and the fynbos or fynbosch of South Africa. The maquis includes *Cistus, Arbutus*, heather, gorse and broom, besides the plants mentioned above in the tall shrub-layer.

(ii) *Garrigue*. Scrubland, with lower, rounded, hard-leaved shrubs (50 cm), is commonly found in dry locations, especially on limestones. It is often a degraded form of maquis (see Figure 5.9). This low-scrub community is known in Europe as garrigue (see Plate 5.3). Many of the species of the taller maquis may still be present, together with such plants as thyme, sage, lavender and rosemary. These plants contain resinous, aromatic oils and have tiny, leathery, hair-covered leaves. This 'rock heath' vegetation is often beautifully colourful in the spring because there are also tuberous plants such as tulip, crocuses, irises and garlic.

(iii) *Arid grasslands*. In very dry areas and also where the vegetation has been seriously affected by burning and grazing, there are drought-resistant grasses, clovers, deep-rooted herbaceous perennials and thistle-like plants. Some plants, such as the asphodel, have underground storage organs and these plants are characteristic of very degraded and impoverished soils.

ASSIGNMENTS
1. *Refer to Figures 2.7, 2.8, 5.3, 5.6, 5.7 and Plates 5.1, 5.2.*
   Compare the boreal coniferous forest biome and the temperate deciduous forest biome in terms of: (i) distribution and modification by humans; (ii) climatic conditions; (iii) structure and composition of species (including animals); (iv) associated soils.

Figure 5.10 Climatic diagrams summarising conditions typical of (a) tropical deciduous forest and (b) tropical semi-evergreen forest biomes in India (Source: Walter 1973)

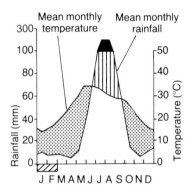

(a) Tropical deciduous forest

Jaipur
Altitude: 434 m
Mean annual temperature: 24.8°C
Mean annual rainfall: 610 mm

(b) Tropical semi-evergreen forest

Belgaum
Altitude: 781 m
Mean annual temperature: 23.2°C
Mean annual rainfall: 1293 mm

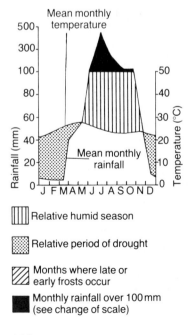

|||| Relative humid season

▒ Relative period of drought

▨ Months where late or early frosts occur

■ Monthly rainfall over 100 mm (see change of scale)

2. (a) *Using Figures 5.3, 5.9 and Plate 5.3, describe the distribution and composition of the mixed evergreen forests of the temperate zone.*

(d) *Using Figure 5.8, describe the climatic conditions associated with this biome.*

(c) *Illustrate how the plant species of this biome adapt to: (i) climatic drought; (ii) human impact.*

C. Tropical Forest Biomes

Introduction

Three main types of tropical forest are found within the tropical zone: tropical rain, tropical deciduous and tropical scrub forest (see Figure 5.3). In the tropics, rainfall rather than temperatures (which are uniformly high) determines the nature and distribution of vegetation. For instance, in the lowlands, where rainfall is less than 500 mm per year and there is a long dry season, desert and various scrub and thorn forests occus. Broad-leaf deciduous forest (e.g. in eastern Brazil, India, Burma, Thailand) occurs where there is 500–2000 mm rainfall per year and a dry season of 2–6 months. A dry month is one with less than 100 mm rainfall. In the wetter parts of this climatic range, broad-leaf, semi-evergreen forest may be found. In this type of forest, the taller trees lose their leaves in the dry season but many of the lower ones keep theirs (see Figure 5.10). In regions with more than 2000 mm of well-distributed rainfall, tropical evergreen rain forest is found (see Figure 5.11).

1. Tropical rain forest biome

(a) *Distribution*

The main areas where tropical rain forest is found include: (1) the Amazon Basin of South America with extensions northward into Central America and southward along the Brazilian coast; (2) the Zaire (formerly Congo) Basin in Africa and parts of West Africa; (3) a large fragmented area in Indo-Malesia, stretching from India and Assam through Indonesia to the north coast of Australia.

(b) *Climate*

Tropical rain forests occur in the humid tropics where rainfall is plentiful and well distributed throughout the year. As shown by the diagrams in Figure 5.11, most of the area covered by this biome has an annual rainfall in excess of 2000 mm; even 3000 mm and more is not uncommon. Over 100 mm of rain in most months is characteristic but there are often dry spells, lasting one or two months, where average monthly rainfall falls below 100 mm (e.g. in Colombo). The mean annual temperature of the rain-forest region is about 27 °C, ranging on a monthly average basis from

24–28 °C. Seasonal variations in temperature are thus small and less than the daily temperature swing which ranges between 6–12 °C.

(c) Vegetation response

The forest trees are large, at least 30–50 metres in height, and are evergreen in overall appearance. There is a great variety of trees, with typically 40–100 species per hectare. One 23-hectare plot in the West Malaysian rain forest contained 375 species. The popular idea of the tropical jungle – thick, steamy and impenetrable – is found only in those areas that humans have at some time cleared, especially along river margins. Figure 5.12 shows an idealised transect across a climax tropical rain forest. The shrub- and ground-layers have relatively few species because of the heavy shade cast by the dense canopy layers above. Most of the plant (and animal) species of these forests are in the canopy where there is plenty of light, unlike in temperate forests where the greater number of species is usually in the undergrowth. The crowns of the trees are covered in *epiphytes*, i.e. plants that use the trees only for support but are not parasites. The system is also rich in large, woody climbers, or lianas, which are rooted in the ground but are carried by the trees into the canopy where they have their leaves and flowers. Altogether the tropical rain forest supports the largest number of plant species of any biome. (See Plate 5.4.)

Figure 5.11 Climatic diagrams summarising the tropical rain forest environment in: (a) Sri Lanka; (b) New Guinea (Source: Walter 1973)

(a)

Colombo Sri Lanka
Altitude: 7 m
Mean annual temperature: 26.6 °C
Mean annual rainfall: 2370 mm

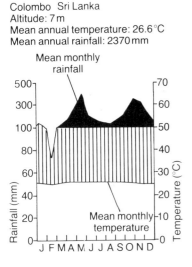

(b)

Suva New Guinea
Altitude: 6 m
Mean annual temperature: 25.6 °C
Mean annual rainfall: 2926 mm

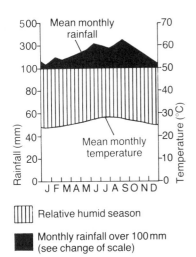

Plate 5.4 Tropical rain forest, Sarawak, East Malaysia. The dense understorey of young trees, climbers and shrubs suggests that this forest is a secondary growth forest. The original forest has been disturbed and cut (note the fallen log), allowing light to penetrate to the ground surface. (Photograph: Barnaby's Picture Library/K. N. Radford)

Figure 5.12 Structure of evergreen tropical rain forest showing the stratification and variations in the shape of the tree crown (Source: Richards, 1979)

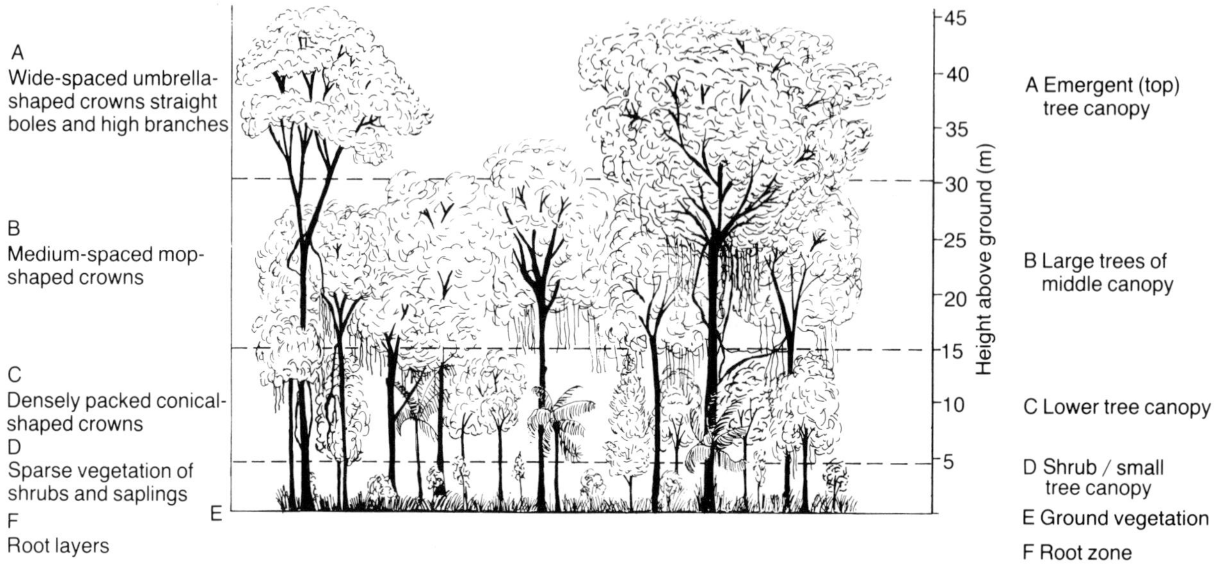

A
Wide-spaced umbrella-shaped crowns straight boles and high branches

B
Medium-spaced mop-shaped crowns

C
Densely packed conical-shaped crowns
D
Sparse vegetation of shrubs and saplings
F
Root layers

E

Height above ground (m)

A Emergent (top) tree canopy

B Large trees of middle canopy

C Lower tree canopy

D Shrub / small tree canopy

E Ground vegetation

F Root zone

(d)   *Animals*

The tropical rain forest biome also contains the greatest variety of animal life (mammals, birds, insects, micro-organisms) of any biome. This is because it offers such a rich selection of plant species and food resources, and such constant environmental conditions throughout the year. There are lots of canopy-dwelling birds which have many different diets, eating seeds, fruit, buds, nectar or insects. Many of the mammals are adapted to living in trees (e.g. sloths, monkeys, ant-eaters). There are also a fair number of ground-living animals, including deer and rodents, which depend on seeds and fruit falling from the canopy layers. Finally, amphibia (semi-aquatic species, e.g. frogs) and reptiles (especially snakes) are important predators of small animals.

(e)   *Soils*

The soils of the humid tropics have three main features (see Figure 5.13). (1) They are intensely weathered, often down to ten metres or more, with high accumulations of strongly altered kaolinitic clay. (2) There is intense leaching and high acidity. However, the high temperatures of these regions result in the movement of silica, rather than iron and aluminium which remain concentrated in the upper soil horizons. These iron-rich, red soils (see Plate 5.5 on page 76) are known as *latosols* (see Figure 5.13b). (3) They are relatively poor in organic matter because of the very high rates of decay in tropical climates.

Some latosols contain subsoil layers exceptionally rich in concentrations of iron and clay known as *plinthite*. When these plinthitic layers

Figure 5.13 Tropical latosol showing: (a) weathering and leaching processes; (b) profile characteristics

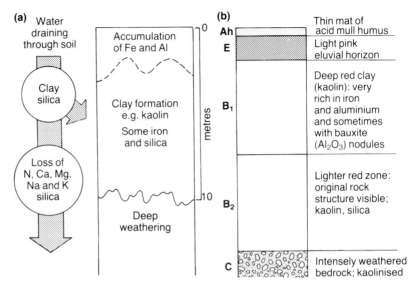

dry out and harden, at or near the surface, they are termed *laterites* (lateritic means brick-like).

While latosols are the characteristic soil of the humid and sub-humid tropics, red-yellow (podzolic) soils (see Figure 5.1) occur in sub-tropical and warm temperate regions (south-east USA, Mediterranean Basin, south-east China). These soils are similar to the latosols, but lower temperatures encourage more silica to remain in the soil and higher levels of organic matter to accumulate. (See Plate 5.6 on page 76.)

2. Soil-vegetation catena in the Maya Mountains, Belize

We now look at a close study which was carried out on the relationships between soil and vegetation and slopes in the humid tropics. Working on tropical hill-slopes, P. A. Furley identified a number of interesting soil catenas in the Northern Maya mountain region of Belize, Central America. One of these soil catenas is shown in Figure 5.14. It is located over a slope of hardened, or metamorphic, mudstones (phyllites) and shows the following variations.

(i) *Soil and vegetation.* There is more water available, by surface run off and throughflow, to the soil and vegetation towards the foot of the slope than at the top. The better supply of water increases plant growth and variety. The dominant vegetation over the drier summit and upper slopes is savanna grassland and shrubs with scattered pine. This community changes rapidly down the slope to a shrub-oak section on the main slope and then to a large, stratified, broad-leaf rain forest on the wetter soils of the lower main and concave slopes.

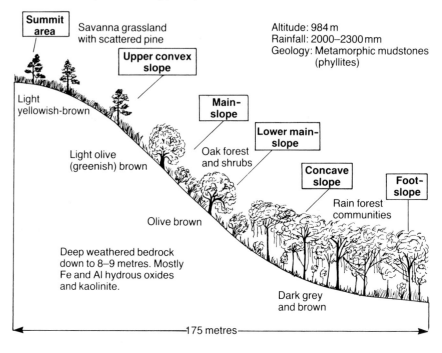

Figure 5.14 Soil-vegetation sequence on Cooma Cairn, Northern Maya Mountains, Belize, Central America (Source: Furley, 1974)

(ii) *Levels of soil nutrients*. The higher productivity and the greater biomass of the vegetation on the lower slopes produce greater incorporation of soil organic matter and, presumably, more efficient means of circulating nutrients. This in turn leads to an increase in many other soil values including soil pH and exchangeable calcium. Soluble bases such as calcium may also be concentrated on the lower slopes, however, by downwash and leaching from upper slopes. There is evidence of a narrow but distinctive foot-slope section where many soil properties decline, especially soil pH and calcium. This may be caused by the trapping of soil bases further up the slope by the tropical rain forest, or by some interference from alluvial deposits from a neighbouring stream.

(iii) *Soil colour*. Soil colour also changes down the slope, from light-yellowish-brown through progressively 'darker' colours to dark grey and brown on the lower slopes. There are two reasons for this. As more organic matter is incorporated into the soil down the slope, so the soil becomes darker. Also, changes in oxidation/reduction processes (see page 35) between the hill-top and the toe-slope darken the soil. The high amount of oxygen characteristic of the well-drained upper slopes encourages light red, non-hydrated ferric iron oxides to form in the soil. Lower, less well-drained, and wetter slopes are characterised by increasing degrees of hydration of iron within the soil; iron compounds are reduced, giving the typical grey, dark brown colours of ferrous iron.

3. Destruction of forests

Today tropical rain forests are being rapidly cleared for a variety of reasons by a variety of people. Because of farming, ranching, logging and the demand for fuel, tropical rain forests are being cleared at a rate of about 110 000–120 000 km² per year. That is equivalent to clearing the size of a football field every second! Of the original area 5000 years ago of 16 million km² of rain forest, only about two-fifths now remains (6.5 million km²). 45% of the present forest is in Latin America, 36% is in Africa, while the remaining 19% is in Indonesia. If this rate of destruction continues for another 20–30 years, we shall see the death of this biome.

(a) Means of destruction

(i) *Cultivation*. This has so far been responsible for about 80% of the total loss of forest. It has been estimated that in the mid-1970s there were at least 150 million peasant cultivators throughout the tropics occupying some 2 million km² of rain forest (one-third of the present total). By clearing relatively small plots of land by fire and axe, these 'shifting' (ever-moving) cultivators convert the forest to agricultural use at a rate of 80 000–90 000 km² per year. 85% of this reduction has taken place in the more highly populated region of Indonesia. In short, farming is reducing or converting rain forest every year by 1.5–2% of its area.

(ii) *Ranching*. Between 1950 and 1975, the area of pasture-land in Central America doubled, mainly because forest was destroyed for this purpose. Between 1966 and 1978, some 80 000 km² of tropical rain forest in Brazil (one-fifth of the total) disappeared when it was converted into 300 very large cattle ranches with some 6 million cattle (see Plate 5.7). Timber was burnt rather than being used or sold. Such tropical pasture (as with the plots of the shifting cultivators) can be easily exhausted and so the cattle are moved from region to region as soon as more forest is destroyed.

(iii) *Logging*. Logging represents a growing menace to the rain forests. Multi-national timber companies have already cleared great areas of forest, especially in Indonesia (three-quarters of the total), but they are increasing the destruction in Amazonia and in central Africa.

(iv) *Fuel*. Not much tropical moist forest is used for fuel at the moment. Most fuel comes from the clearance of secondary (i.e. re-grown) deciduous forest and scrubland in surrounding drier areas. However, with charcoal being used much more, especially in urban areas, distant forests are beginning to be affected: forests far from Bangkok are being converted into charcoal, as are those of north-west Kenya for sale in Nairobi.

(b) Consequences of forest removal

There are a number of reasons why we should mourn the loss and conversion of the tropical rain forest. (1) There is an obvious concern

Plate 5.7 Clearance of tropical rain forest in Rondonia, Brazil. The rain forest has simply been burnt to create poor pasture land for cattle ranching (Photograph: Marios Santilli/Panos Pictures).

about the loss of the rich and unique variety of plant and animal life of the biome. (2) The rain forest has a very large 'genetic pool' of plants and animals. This genetic variety, which is neither fully understood nor appreciated, must be conserved for future use. Humans get important economic, medical and scientific benefits from using tropical plants and animals. (3) The biome provides wild areas for recreation, enjoyment and education. For many developing countries it provides local income and foreign exchange through tourism. (4) The tropical rain forests need to be conserved to maintain climatic and ecological stability. It has been claimed that clearing the forests in this biome increases local aridity because of decreased transpiration rates after the forest has been destroyed. Rainfall declines and becomes less frequent when this source of atmospheric moisture is removed. In addition, it has been suggested that large-scale removal of forests may increase the concentration of carbon dioxide and decrease the level of oxygen in the global atmosphere. Rapid rates of forest clearance will increase levels of carbon dioxide through burning and the decomposition of the vegetation. Higher levels of carbon dioxide may produce warming of the earth's surface. Such warming may lead to the melting of glaciers and ice caps, resulting in devastation from rising sea-levels and flooding (see O'Hare and Sweeney, 1986). Removing forests in the tropics also encourages a variety of harmful ecological effects, including *laterisation* and soil exhaustion, desertification and increased rates of soil leaching and erosion (see pages 146, 147).

ASSIGNMENTS

1. (a) *Name three types of tropical forest.*
   (b) *Using Figures 5.3 and 5.11, describe and account for the distribution of tropical rain forest.*

(c) *Using Figure 5.12 and Plate 5.4 describe (i) the structure or stratification and (ii) the composition of species of the tropical rain forest, noting any important relationships between the two.*

(d) *With reference to Figure 5.13 and Plate 5.5, list the chief features of tropical latosols. What roles do climate and vegetation play in their formation?*

(e) *Distinguish between latosols and laterites.*

2. (a) *Describe the catenary variations in soil and vegetation shown in Figure 5.14.*

(b) *Examine the main processes responsible for these variations.*

D. Grassland Biomes

1. Temperate grassland biome

(a) *Distribution and climate*

This biome includes large, 'natural' grassland areas such as the prairies of North America, the steppes of Eurasia, the pampas of South America and the veld of South Africa (see Figure 5.3). The dominant plants in all of them are the grasses, the most widespread and successful group of land plants. These grasslands are very ancient and are considered to be natural climax formations determined by soil and climate. Large areas of grassland still remain despite widespread transformation by cultivation, especially for grain crops such as wheat and maize (see Plate 5.8).

The temperate grassland of the mid-continental interiors occurs in regions where rainfall is midway between that of desert and that of temperate forest. Average precipitation is 250–750 mm and there is a fairly long dry season (see Figure 5.15). Within these grasslands a distinction is often made between that of tall-grass prairie (about 1 metre high) in

Figure 5.15 Characteristic climate of the sub-humid temperate grasslands. The diagram shows a wide range of temperature between summer and winter, a summer maximum rainfall and a tendency for droughts to occur in early and (especially) in late summer. (Source: Walter, 1973)

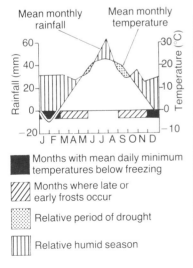

Odessa
Altitude: 70 m
Mean annual temperature: 9.9 °C
Mean annual rainfall: 392 mm

Plate 5.8 Prairie grassland biome, North America. Much of the prairie grassland of central North America has been converted to grain production, shown by the wheat field in the foreground. Prairie grassland with scattered woodland is found on the steeper marginal slopes. (Photograph: J. Tivy)

the more humid regions of the biome, and short-grass prairie (about 50 cm high) in more arid sections.

Other areas of temperate grassland exist in Britain and Europe, for instance. However, these are created by farmers out of the destruction of forests and woodlands.

(b) Soils

(i) *Chernozems*. The characteristic soil of the tall-grass, temperate grass-land biome is known as a *chernozem* (see Figure 5.16a, and Plate 5.9 on page 76). Chernozems are black soils typically developed from parent materials which are rich in loess, a wind-blown deposit of fine, silty, calcareous material. The thick, surface A-horizon, which may be up to 1 metre in depth, is made up of very dark humus with a fine granular or crumb structure, resulting from the activities of earthworms and other fauna. This horizon grades gradually into a lighter coloured, mineral-organic A/C-horizon which often has burrows (krotovinas) of small rodents. This horizon rests directly on the C-horizon rich in secretions of calcium carbonate (Cca).

Chernozems show how climate and vegetation affect soil formation. The richness of humus in these soils is a result of a good relationship between input and output: the rapid growth of tall grassland in the hot, moist spring and early summer encourages large organic inputs in the form of leaf- and root-decay; the dryness of mid- and late summer and the low temperatures of winter restrict the activity of micro-organisms and thus the loss of organic remains by decomposition.

Light rainfall and high rates of evaporation (i.e. low P/Pet ratios)

Figure 5.16 Grassland soils showing: (a) chernozem associated with tall-grass prairie; (b) chestnut soil associated with short-grass prairie

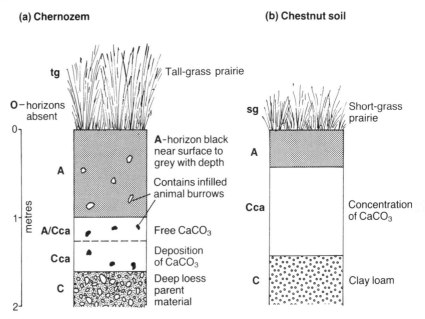

(a) Chernozem

(b) Chestnut soil

result in only a mild degree of leaching and so the upper horizons are neutral or only slightly acid. The absence of strong leaching leads to calcareous substances being concentrated in the lower horizons and nutrients being held within the rooting zone. Mull-humus forms in this neutral, calcium-rich environment and is thoroughly mixed with the soil mineral matter by very active soil organisms.

(ii) *Chestnut-brown soils and degraded chernozems.* Not all of the area designated as temperate grassland in Figure 5.3 is associated with black chernozems. In North America, for instance, *chestnut-brown soils* are found to the west and south-west, where there are lower rates of rainfall and higher rates of evaporation (see Figure 5.1). As shown in Figure 5.16b, these soils occur under a shorter, less productive grassland. They have, as a result, a shallower (30–50 cm) organic A-horizon and accumulations of calcium carbonate closer to the surface (50–100 cm).

On the more humid margins of the temperate grasslands (to the east in North America), close to the temperate deciduous forest biome, *degraded chernozems* or *prairie* soils occur. These soils have features midway between the chernozems of the central grasslands and the brown podzolic soils of the forested areas (see Figure 5.17). They lack the strongly eluviated (Ea) and illuviated (B) horizons of the forest soil. They are, however, more leached and acid than the chernozems, and have enough downward translocation and deposition of substances in the subsoil for a B-horizon to form. Being transitional between the chernozems and the brown podzolic soils, they have intermediate accumulations of organic remains in the A-horizon.

Figure 5.17 Typical soil changes in North America along a west-to-east transect, from densely vegetated prairie grassland (deep black topsoil) to forest (shallow topsoil). Transitional, or degraded chernozems (prairie soils), occur between forest and grassland. They are found: (i) where woodland has encroached on the grassland; (ii) where woodland has been removed, allowing grassland to invade. (R = unweathered bedrock, e.g. loess.)

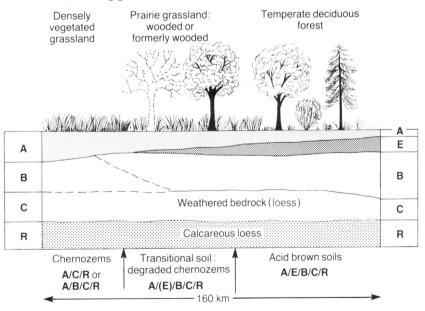

2. Tropical grassland biome

(a) Distribution and origin

Tropical grasslands are commonly referred to as *savanna*. Savanna is a term applied to any tropical vegetation ranging from pure grassland to woodland with much grass. As shown in Figure 5.3, this biome covers a wide belt on either side of the Equator between the Tropics of Cancer and Capricorn. The climate is always very warm and there is a long dry season. However, climatic conditions alone do not influence the distribution of this biome. Total rainfall, for instance, tends to range from 500 mm to over 2000 mm per year. Thus many tropical savanna areas have enough rainfall to support some type of deciduous and thorn-scrub forest vegetation. Many savannas, especially in Africa, are probably brought about by humans who have created them from previous woodland and forest cover by cutting, fire and grazing (see Plate 5.10).

(b) Vegetation, animals and soils

In spite of high rainfall, the plant species of the savanna have drought-resisting features. The grass can grow to 3.5 metres and so is much taller than that of temperate grassland. There is often a great variety of trees which also show drought-resisting, or xeromorphic, adaptations, e.g. Acacias, baobab trees.

The dominant animals are large, grazing mammals (e.g. zebra, wildebeest, antelope, elephant). The African savanna has the greatest variety.

Plate 5.10 Bush savanna, western Tsavo, Kenya, showing a variety of animals, e.g. elephants, zebra and antelope. (Photograph: Topham/C. Osborne)

126

Burrowing rodents are also characteristic. Large carnivores, such as lions and hyenas, are the natural predators of the grazing animals.

The characteristic soil of the savanna grasslands is similar to that of the tropical rain forest, i.e. red, iron-rich latosol. However, in view of lower rainfall and a prolonged dry season, the soil is generally less intensely weathered and leached. Silica is not so easily moved and accumulations of organic matter are slightly higher.

ASSIGNMENTS

1. (a) *Refer to Figure 5.3 and Plate 5.8. Describe the nature and distribution of the temperate grassland biome.*

   (b) *Refer to Figure 5.15. What role does climate play in the formation of the temperate grassland of mid-continental interiors?*

   (c) *Using Figure 5.16 and Plate 5.9, describe the profile features of the characteristic soil of the mid-continental temperate grasslands.*

   (d) *What roles do climate and vegetation play in the development of this soil?*

   (e) *Name and briefly describe two other soils found in the temperate grassland biome.*

2. (a) *What is savanna grassland?*

   (b) *How does savanna grassland differ from temperate grassland with respect to: (i) distribution; (ii) origin and formation; (iii) associated soil types?*

E. Desert and Tundra Biomes

1. Arid desert biomes

(a) Distribution and climate

The arid deserts are defined as areas, apart from the cold tundra regions, where there is less than 250 mm rainfall per year. Their distribution is shown in Figure 5.3. There are both hot and cold arid deserts. The hot arid deserts, such as the Sahara of Africa, the Thar Desert of India and the Great Australian Desert, are located in tropical and sub-tropical areas. They have very high day-time temperatures, often over 50 °C, and low night-time temperatures, below 20 °C, with relatively mild winters (see Figure 5.18). In contrast, the cold arid deserts are located in temperate latitudes, either at the centre of large land masses (e.g. the Gobi Desert of Mongolia) or in the *rain-shadow* of mountain ranges (e.g. Patagonia or the desert of south-west USA). Such cold deserts have very severe winters and long periods of extreme cold.

(b) Vegetation and animal response

Desert vegetation is characteristically very sparse and scattered (see Plates 5.11 and 5.12), since there is not enough moisture available to support a thick, continuous plant cover. Individual plants cope with the scarcity of water in various ways. Some plants have drought-resistant seeds which

Figure 5.18 Climatic characteristics of a hot, arid biome. Temperatures tend to fall below zero at night in the winter period, giving a chance of frost. Despite a peak of rainfall in summer, there is a year-long lack of soil moisture because of high evaporation losses. (Source: Furley & Newey, 1982)

Alice Springs
Altitude: 621 m
Mean annual temperature: 20.9 °C
Mean annual rainfall: 275 mm

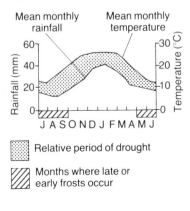

Mean monthly rainfall   Mean monthly temperature

Relative period of drought

Months where late or early frosts occur

remain dormant for long periods but quickly germinate as soon as rain falls. Some have small, thick leaves that are shed in dry periods. Some, such as the cacti, are succulents which store water in their stems. Others have very deep roots and are able to tap underground water supplies.

Desert animals survive the aridity and heat by keeping cool and by preventing loss of water. Many are small enough to hide under stones or in burrows during the intense day-time heat. Certain rodents live in cool burrows, are largely nocturnal (i.e. their main active period is during the night) and pass very little water in their urine. Insects and reptiles, which are common creatures of the desert, have waterproof skins and produce very little urine. Many desert animals (e.g. the camel) can survive for long periods in the desert without needing to consume water on a daily basis.

(c) Desert vegetation: Death Valley, California

In the Sierra Nevada Mountains of south-west USA, there are zones of vegetation which vary according to altitude, with boreal forest on the high summit-slopes and desert vegetation in the valleys. These zones certainly reflect the distribution of climate at various altitudes, e.g. cold, moist uplands and hot, dry valleys in the rain-shadow of the mountains.

The detailed distribution of vegetation within one of these valleys, Death Valley, California, which has less than 50 mm average annual rain-fall, depends on the ground and soil conditions, especially those affecting the quantity and quality of the water-supply. The three main soil/landform conditions are: the central salt pan (playa); the lower, sandy foot-slope; and the upper gravel fans (see Figure 5.19).

Plate 5.11 Cattle, goats, sheep and camels water at El Beshiri oasis, Sudan. Increasing herd sizes or livestock density, and decreasing pasture lands, leads to overgrazing. Here the desert is expanding (i.e. by desertification) as herds have eaten or trampled all vegetation for as far as the eye can see. (Photograph: Mark Edwards/Panos Pictures)

(i) *The central playa*. In enclosed depressions like Death Valley, rain-water containing dissolved chemical elements collects by run off from the surrounding higher ground. As water seeps down, it helps to raise groundwater tables close to the surface, especially where underlying drainage is poor (see Figure 2.5 on page 34). Because of the very high temperatures of this valley, groundwater containing large quantities of salts in solution is drawn upwards to the surface under strong evapor-ation. As the water evaporates, salts are deposited as thick crusts on the surface (see Plate 5.12). This process of *salinisation* (see pages 143–5) is responsible for the high accumulations of salt in the central playa or salt pan of Death Valley. The concentrations of salt are so excessive here that no flowering plants are found.

Note that many arid, desert soils (and not just those in enclosed depressions like Death Valley) are affected to varying degrees by salin-

Figure 5.19 Transect across Death Valley, California, showing the orderly zoning of vegetation, which reflects the availability and salt-content of water for plant growth (Source: Hunt, 1972)

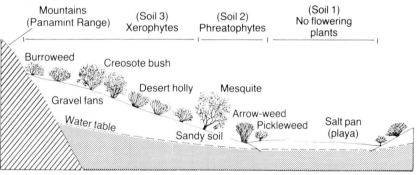

Plate 5.12 Desert scrub vegetation, Nevada Desert, USA. The white areas in this photo-graph are high concentrations of salt on the soil surface. Crusts of salt form in the surface layers of the soil, because of high rates of evaporation, in a process known as salinisation (Photograph: J. Tivy)

129

Figure 5.20 Model illustrating the occurrence and type of sub-surface water in Death Valley. Phreatophytes have their roots in the deeper (phreatic) groundwater and fringewater; the roots of xerophytes do not reach the top of the capillary fringe. (Source: Hunt, 1966)

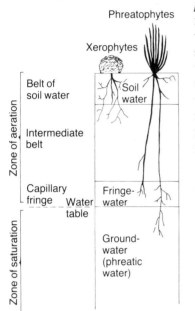

isation. Often irrigation, by supplying additional water to such soils, has worsened the problems of surface alkalinity and salinity by increasing the upward movement of salts from raised water tables.

(ii) *Sandy foot-slope/phreatophyte zone.* Around the edge of the pan is a belt about 1.6 km wide, where the ground is sandy and where the groundwater is reasonably close to the surface. In general, this is where *phreatophytes* occur. These are plants that live on dry ground but whose root-systems extend to the permanent water table (see Figure 5.20). The various phreatophytes are arranged in an olderly way according to the saltiness (salinity) of the groundwater. The most salt-tolerant species is the pickleweed. This plant can tolerate a surface concentration of salt of up to 6% (i.e. twice the concentration of sea water!) and occurs closest to the salt pan. Beyond this is the arrow-weed, which can tolerate as much as 3% concentration of salt (about the same as sea water). Outward from the arrow-weed, and furthest from the salt pan, is the honey mesquite which can grow on a maximum of 0.5% salt content.

(iii) *Gravel fan/xerophyte zone.* Between the sandy ground and the foot of the mountains are gravel fans which are the habitat for *xerophytes* (i.e. drought-resisting plants) such as the creosote bush. Although xerophyte roots may be extensive, they do not extend to the water table which is many tens of metres from the gravel surface. The only water available for plant growth is that from infrequent rainfall, dew, and soil moisture contained in the upper surface gravels (see Figure 5.20). The xerophytes can survive the dryness of the gravels. The reason why they cannot survive in the phraetophyte zone, where groundwater is shallow, may be the salinity of the environment or the fact that their roots are drowned during times of high groundwater. Indeed, the maximum amount of salts in the gravels supporting xerophyte vegetation is where there are communities or stands of desert holly only. At these stands, the salt content may be as high as 1%; where the burroweed occurs, the salts are less than 0.25%.

Different amounts of soil moisture may also explain the distribution of the xerophytes. Soil moisture is possibly greatest at the top of the gravel fans, because these are close to surface run off from the nearby mountains (Panamint Range), and lowest at the base of the fans. This gradient probably determines the location sequence: the burroweed (least xeric or drought-resisting) above the creosote bush above the desert holly (most xeric).

2. Tundra biomes

The word *tundra* means barren land. The tundra biomes are located in Arctic or Alpine regions where no trees grow because environmental conditions are so severe (see Plate 5.13). They are found around the Arctic Circle north of the sub-Arctic coniferous forest (see Figure 5.3). Alpine tundra occurs above the tree line on high mountains, including those in the tropics. We shall look at the Arctic tundra.

(a) Climate

Air temperatures in the Arctic tundra are low the whole year round.

Figure 5.21 The climate of the tundra showing very cool summers, long, very cold winters, and low precipitation with a summer maximum (Source: Walter, 1973)

Fort Yukon
Altitude: 127 m
Mean annual temperature: −6.7 °C
Mean annual rainfall: 172 mm

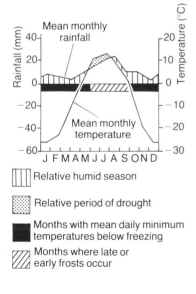

Annual averages are not much above freezing point (0 °C), with six to ten months having monthly temperatures below freezing (see Figure 5.21). Precipitation, much of it in the form of snow, is low and irregular and is seldom over 250–300 mm except in areas near the ocean. During the brief summer, water melts at the soil surface where temperatures may reach 12–15 °C; but underneath there is always a permanent layer of frozen soil called permafrost. There is a very short growing season, often less than 50 days between the spring and autumn frosts.

(b)   Vegetation response

Only plants which can tolerate cold can survive this environment. Floristically the tundra biome is poor in species. There are only about 100 species of higher (i.e. more evolved) plants. The principal plant groups are mosses, lichens, hardy grasses, sedges, dwarf shrubs and herbs (see Plate 5.13). There is, however, great spatial variation in plant composition.

According to latitude, from the high Arctic tundra nearest the poles to the low Arctic tundra close to the Arctic Circle, there is a gradual gradation from mosses and lichens, to sedges and grasses, to shrubby heaths and eventually to grassland and heathland with dwarf trees (e.g. willow, birch). Vegetation also changes at the local scale depending on highly variable conditions of soil and relief. In low-lying, poorly drained, peaty depressions, mosses, grasses and sedges predominate. On slightly higher, better-drained slopes, heathland with dwarf shrubs (heather) and grasses are found above podzols. In sheltered valleys, thickets of dwarf birch and creeping Arctic willow occur, often in association with brown podzolic soils. Finally, on higher exposed plateaux, rock deserts or 'fellfields' are characteristic. These exposed, snow-free areas support only a sparse growth of mosses and lichens over thin, dry, rock-strewn lithosols (see page 45), which are low in organic matter and can hold little water.

(c) Animals

Large herbivores include reindeer, caribou and musk ox. Small herbivores are typified by hares, lemmings and voles. The large herbivores cannot be supported all year by the tundra biome. They migrate in summer from the southern boreal forest to the tundra, feeding on the rapid increase in vegetation growth which takes place during this season. Many birds do likewise, feeding on the large insect populations. Carnivores include Arctic fox, wolves, hawks, falcons and owls.

ASSIGNMENTS

1. (a) *Define and name two types of arid desert.*
    (b) *How do desert plants and animals adapt to heat and the scarcity of water?*
2. (a) *Describe the distribution of vegetation shown in Figure 5.19.*
    (b) *Describe the process of salinisation.*
    (c) *What factors encourage its development?*
    (d) *Examine the relationship between the distribution of vegetation and the physical conditions in Death Valley.*
3. (a) *Characterise the climate of the tundra shown in Figure 5.21.*
    (b) *Describe the general effects of the tundra climate on: (i) vegetation; (ii) soils; (iii) animals.*
    (c) *In the tundra, how do climate, vegetation and soils vary: (i) latitudinally; (ii) locally?*

Key Ideas

A. Introduction

1. At global level, there is a close zonal relationship between climate and soils and vegetation.
2. Biomes, which are large-scale ecosystems, are useful in the study of the relationships between world soils and vegetation and climate.
3. The study of biomes can be helped by using carefully constructed climatic diagrams.

B. Temperate Forest Biomes

1. Three types of forest are found in temperate latitudes: boreal coniferous; cool temperate, broad-leaf deciduous; warm temperate, mixed (evergreen and deciduous) forests.
2. The boreal coniferous biome of the northern cold temperate zone is characterised by a cold continental climate, evergreen conifers and podzolic soils.
3. The temperate deciduous forest biome is characterised by a cool temperate climate, broad-leaf deciduous trees and brown soils.
4. Warm temperate mixed forest of conifers and broad-leaf evergreen trees, in association with red, iron-rich soils, can be found in areas such as the Mediterranean Basin which has mild wet winters and warm dry summers.

5. Mediterranean mixed forest has been converted by humans to tall shrubland (maquis), low scrub (garrigue) and arid grassland.
6. Most plants belonging to Mediterranean vegetation have special features to resist or evade drought.

C. Tropical Forest Biomes

1. There are three main types of tropical forest biome including tropical rain forest, tropical deciduous forest and tropical scrub woodland.
2. Tropical rain forest, the most species-rich and highly stratified biome on earth, is found in the humid tropics where rainfall is plentiful and well distributed throughout the year.
3. Red, iron-rich tropical soils, which are found under tropical forest and grassland, are known as latosols.
4. Latosols are highly weathered, intensively leached and relatively poor in organic remains.
5. Vegetation and soils (including soil colour and nutrient content) vary greatly along slopes in the Maya Mountains, Belize, Central America.
6. The tropical rain forest is being rapidly cleared for farming, ranching, timber and fuel.

D. Grassland Biomes

1. Large areas of 'natural', temperate grassland are found in mid-continental interiors, where rainfall is midway between those of forest and desert.
2. The characteristic soil of these grassland areas is the chernozem.
3. Chernozems are mildly leached, black, humus-rich soils, typically found above wind-blown deposits of loess.
4. Chestnut-coloured soils occur under less productive and shorter grasslands, in the transition zone between grassland and desert.
5. Degraded (leached) chernozems or prairie soils are found in the more humid parts of the biome, in the transition zone between grassland and temperate forest.
6. Tropical grassland or savanna is typically found in areas which are always very warm and have a long dry season.
7. Nevertheless many tropical grasslands are located in areas which are moist enough to support forest. They have been created from previous woodland cover by forest clearance.
8. The typical soil of the savannas is the latosol.

E. Desert and Tundra Biomes

1. Arid deserts, whether hot (e.g. Sahara) or cold (e.g. the Gobi), are defined as areas (excluding tundra regions) with less than 250 mm of rainfall.
2. Desert plants and animals have a range of adaptations which allow them to keep cool and to reduce loss of water.
3. Desert vegetation in Death Valley, California, depends on the

ground and soil conditions, especially those affecting the quantity and quality of the water-supply.

4. Drought-resisting xerophytes are found on the dry, upper gravel fans, deep-rooted phreatophytes on the lower sandy slopes, while there is no vegetation in the central salt pan.

5. The treeless Arctic plains of the tundra are characterised by extreme cold and low and irregular precipitation, much of it in the form of snow.

6. The vegetation of the tundra biome is species-poor, consisting of mosses and lichens, hardy grasses and sedges, dwarf shrubs and herbs.

7. This vegetation is spatially varied depending on local soil, relief and hydrological conditions.

8. Many soils in the tundra are shallow, peaty and waterlogged and lie above a permanently frozen layer called permafrost.

Additional Activities

1. (a) Describe the patterns of soil, climate and vegetation shown in Figures 5.1–5.3.
   (b) Using Figures 5.22 and 5.4, examine the relationship between world climate and vegetation.
   (c) Examine the relationship between global climate and soil type shown in Figures 5.23 and 5.4.
   (d) Using Table 5.1 as a guide, now summarise the links between soils, vegetation and climate at the world scale.

2. (a) Describe the patterns of (i) average rainfall, (ii) surface deposits and (iii) vegetation shown in Figure 5.24a–c.
   (b) Describe the distribution of soils shown in 5.24d.
   (c) Describe the nature of the three soil types in Figure 5.17.
   (d) Using your results so far, explain the distribution of soils in North America.

3. Refer to Tables 5.2 and 5.3.
   (a) Describe the three plant communities of Barrow tundra in terms of: (i) plant composition, (ii) size; (iii) productivity.
   (b) Describe the physical habitats associated with each plant community.
   (c) Explain the link between: (i) snow and thaw depth; (ii) content of soil moisture and accumulation of organic matter; (iii) content of soil nutrients and soil acidity.
   (d) Carefully examine the relationship between each plant community and its physical habitat.

Table 5.1 Summary of the global relationships between soils, climate and vegetation. Note that leaching, acidity and humidity increase from desert areas, while average temperature and rate of decomposition of humus increase from cold tundra areas.

Soil type	Soil characteristics	Climate	Vegetation	
Tundra	Peat; subsoil permanently frozen	Perpetually cold	Bog and wet moor	
Podsol	Acid raw humus; topsoil leached of bases and sesquioxides	Cold, moist winter; mild summer rain evenly distributed	Coniferous forest or heathland	leaching of bases and sesquioxides
Brown earth	'Mild' humus, or mull. Some leaching of bases. Surface weakly acid	Moist temperate, or maritime	Deciduous forest	
Chernozem (black earth)	No raw humus, but deep humus horizon in the soil. Surface neutral; little leaching	Continental; cold winter; hot summer; moderate rainfall 400–500 mm	Humid steppe; perennial grasses	
Chestnut-coloured soil	Less humus than in chernozem. Soil slightly alkaline; some accumulation of salts near the surface	Continental, with low rainfall less than 400 mm	Dry steppe; drought-resisting shrubs	very little leaching
Saline and alkali soil	Strongly alkaline, and/or accumulation of salts in upper horizons. Low humus content.	Arid; less than 250 mm rainfall	Salt-tolerant plants, mosses, lichens	
Sub-humid tropical soil	Rather acid. Red or reddish grey. Very little humus. Some concentration of sesquioxides at surface.	Hot; alternate wet and dry seasons	Tropical deciduous forest or grassland savanna	leaching of bases and silica
Moist tropical soil	Acid; brick red; practically no humus. Leached of bases and much silica. The surface layer consists largely of hydrated sesquioxides and quartz	Hot with high rainfall.	Tropical rain forest	

— Desert

1 Leaching   2 Acidity   3 Humidity   4 Average temperature   5 Rate of decomposition of humus
(Values increase in direction of arrows.)

1,2,3   4,5

135

Figure 5.22   Climatic relationships of six major biomes (Source: Hammond, 1972)

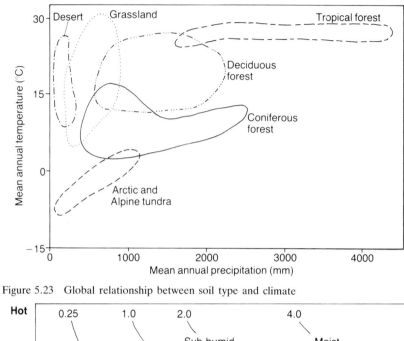

Figure 5.23   Global relationship between soil type and climate

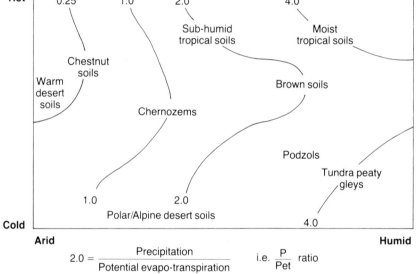

$$2.0 = \frac{\text{Precipitation}}{\text{Potential evapo-transpiration}} \quad \text{i.e. } \frac{P}{Pet} \text{ ratio}$$

Table 5.2   Characteristics of plant communities in relation to habitat on a raised mound, Barrow tundra, Alaska

Type of habitat (raised mound)	Vegetation biomass (g/m²)						Productivity (g/m²/yr)
	Herbs	Woody shrubs	Grasses	Lichens	Mosses	Total	
1 Exposed high central dome	4.5	5.4	9.4	14.0	15.6	48.9	18.1
2 Well-drained middle slope	5.2	31.0	4.1	37.3	7.5	85.1	25.3
3 Wet, low hollow	1.8	0	44.8	0.2	36.7	83.5	43.9

Source: Webber, 1978

Table 5.3 Habitat conditions in relation to plant communities (Barrow tundra, Alaska)

Habitat conditions	Habitat type		
	1	2	3
Soil moisture (%)	77	64	379
Snow depth (cm)	0	4	36
Thaw depth (cm)	37	54	32
Soil pH	3.9	5.3	4.2
Organic matter (%)	24	12	73
Organic horizon (cm)	4.4	3.7	18.2
Soil phosphate (mg/gdw soil)	15.5	21.5	4.5

Source: Webber, 1978

Figure 5.24 North American environment showing: (a) average annual rainfall; (b) surface deposits; (c) vegetation belts; (d) major soil groups (Source: Hunt, 1972)

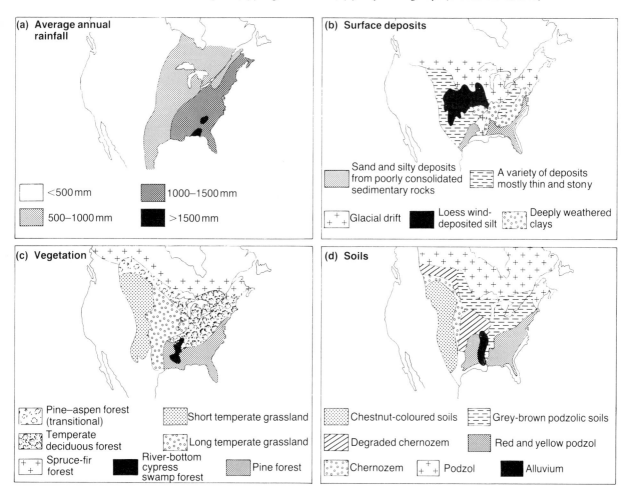

The Global Ecosystems: Process and Response

Introduction

As well as being different in physical appearance, the various biomes also work or function in contrasting ways. This chapter examines some of these contrasts with respect to three processes: (1) plant productivity and growth rate; (2) storage and circulation of reserves of nutrients; (3) impact of usage and disturbance by humans.

A. Plant Productivity of the World's Biomes

1. Distribution of productivity

As described in Chapter 4, the rate at which vegetation grows is termed net primary productivity. This is the net amount of organic material fixed, in photosynthesis, by plants in a unit of time: the material available for harvest or consumption by animals and for decomposition by micro-organisms. Net primary productivity is expressed as a weight of organic matter produced per unit area per unit time, e.g. $kg/m^2/yr$. As shown in Table 6.1, the lowest producers are the tundra and tropical desert biomes with annual rates of $0.14 kg/m^2/yr$ and $0.09 kg/m^2/yr$ respectively. Amongst the forests, which are generally high producers, we can trace a gradual increase in productivity from boreal forest ($0.8 kg/m^2/yr$) to the tropical rain forest where average rates reach $2.2 kg/m^2/yr$.

Grasslands are generally less productive than the forest biomes. Temperate grasslands grow only about half as fast as the temperate deciduous forests. Tropical grasslands, while less productive than most forest biomes, do have a higher growth rate than the slowest-growing forest biome, the boreal forest.

The position of agricultural land is of interest, with a world average of about $0.65 kg/m^2/yr$. This exceeds the average for desert and tundra but falls below many forests and some grasslands.

2. Explaining variations in growth rates

Forests, with their large biomasses (see Table 6.1) generally produce the greatest amount of organic tissue in the course of the year because their large physical structures are able to trap and fix a lot of solar energy. The length and nature of the growing season (the period during which plants

Table 6.1 Net primary production and plant biomass for selected world biomes

Ecosystem type	i Area (10^6 km²)	ii Net primary productivity per unit area (kg/m²/yr)	iii World net primary production (10⁹t/yr) (i.e. i × ii)	iv Biomass per unit area (kg/m²)	v World biomass (10⁹t) (i.e. i × iv)
		Mean		Mean	
Tropical rain forest	17.0	2.2	37.4	45	765
Tropical seasonal forest	7.5	1.6	12.0	35	260
Temperate evergreen forest	5.0	1.3	6.5	35	175
Temperate deciduous forest	7.0	1.2	8.4	30	210
Boreal forest	12.0	0.800	9.6	20	240
Woodland and shrubland	8.5	0.700	6.0	6	51
Savanna	15.0	0.900	13.5	4	60
Temperate grassland	9.0	0.600	5.4	1.6	14
Tundra and Alpine	8.0	0.140	1.1	0.6	5
Desert and semi-desert scrub	18.0	0.090	1.6	0.7	13
Extreme desert, rock, sand and ice	24.0	0.003	0.07	0.02	0.5
Cultivated land	14.0	0.650	9.1	1	14
Swamp and marsh	2.0	2.0	4.0	15	30

Source: Whittaker, 1975

are able to grow) is also important. A year-round growing season, with no frost or drought and with plenty of moisture and heat, enables the tropical rain forest to produce the greatest amount of organic material annually. The tropical rain forest is more productive than tropical deciduous forest, which is hampered by a winter dry period when little growth occurs. Temperate deciduous forest is not as productive as tropical deciduous forest because of the smaller amounts of energy (heat) available for growth (*cf.* Figure 5.7 and 5.10 on pages 112, 116). Nevertheless temperate deciduous forest is more productive than boreal coniferous forest because of a longer, more favourable growth period (*cf.* Figures 5.6 and 5.7 on pages 111, 112).

While the low productivity of the hot deserts can be explained by water shortage, a variety of factors limits the growth rate of vegetation in the tundra: a short growing season (often less than 50 days), cool summer temperatures, a lack of soil nutrients (especially phosphorus and nitrogen), and an unstable, often waterlogged soil.

The relatively low average productivity of agricultural ecosystems is caused by a limited growing season and by the smallness of its biomass. Most crops are annuals and so are in the ground for only a short length of time (about four months in the UK). There is often a large area of open space between crops, especially when young. These conditions limit the ability of cropland to make full use of photosynthetic energy whenever it is available.

ASSIGNMENTS

1. (a) Refer to Table 6.1. Describe the distribution of plant productivity within the major land biomes.

(b) *Examine the factors affecting the productivity of the principal biomes.*

B. Nutrient Cycling in Forest Biomes

In this section the circulation of nutrients in three forest ecosystems (i.e. temperate deciduous, boreal coniferous and tropical rain forest) are compared. Differences in (i) nutrient storage and (ii) nutrient cycling are examined. It may be useful to recall Chapter 4, Section B.2 on nutrient cycling before reading on.

1. Nutrient stores

In the boreal coniferous forest biome, nutrients become 'locked up' in large amounts of surface litter (see Figure 6.1a). The litter compartment is often the main reserve of nutrients in the boreal forest. Relatively few nutrients are contained within the soil (podsol) compartment, where intense leaching is responsible for removing bases. By contrast, there is a much more even distribution of elements in the temperate deciduous forest biome, where both litter and soil (brown soil) are well supplied (Figure 6.1b). Only small quantities of litter accumulate on the floor of the tropical rain forest (Figure 6.1c) because plant debris is rapidly decomposed under hot, wet conditions. The relatively great age of many tropical soils (latosols), coupled with high rates of weathering and leaching, produce deep but generally infertile soils, lacking in nutrients. As a result, the living forest biomass contains the bulk of the reserve of nutrients of this ecosystem. One recent study calculated that 92% of the magnesium, 90% of the potassium, 74% of the calcium and almost 100% of the nitrogen was contained in the plant biomass.

2. Nutrient exchange

(a) *Deciduous* v. *coniferous woodland*

Comparisons of deciduous and coniferous woodland ecosystems consistently show that the amount and rate of nutrient circulation are greater within the former system. This is demonstrated in Figure 6.2, which shows nutrient cycling in hardwood deciduous (European beech) and softwood coniferous (Scots pine) forest. Four nutrients, i.e. potassium (K), calcium (Ca), nitrogen (N) and phosphorus (P), are used to show patterns of circulation.

(i) *Annual uptake from soil.* Although the deciduous forest uses only slightly greater amounts of nitrogen from the soil each year than coniferous forest (50 kg/ha v. 45 kg/ha), hardwood forest demands much more of the other three elements than softwood forest. For instance, the beech forest uses more than twice as much potassium and three times as much calcium and phosphorus as the Scots pine forest.

Figure 6.1 Nutrient circulation, input and loss in three forest biomes. Circle size illustrates the amount of nutrients stored in the compartment. Arrow-width shows nutrient flow as a percentage of the nutrients stored in the source compartment. The diagrams show idealised zonal forest ecosystems, e.g. climatic climax systems in equilibrium with environment (Source: Gersmehl, 1976)

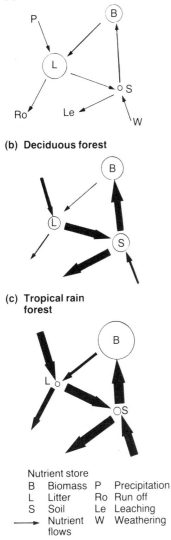

(a) Boreal forest

(b) Deciduous forest

(c) Tropical rain forest

	Nutrient store		
B	Biomass	P	Precipitation
L	Litter	Ro	Run off
S	Soil	Le	Leaching
→	Nutrient flows	W	Weathering

Figure 6.2 Cycling of four nutrients in temperate deciduous (European beech) and boreal coniferous (Scots pine) forests. Units are kg/ha/yr. (Source: Duvigneaud and Denaeyer-De Smet, 1970)

European beech
(*Fagus sylvatica*)

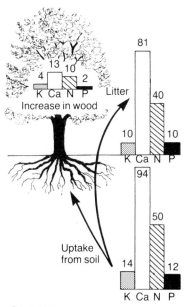

Scots pine
(*Pinus sylvestris*)

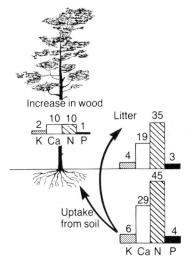

(ii) *Retention in new biomass*. Figure 6.2 shows that a relatively small amount of the total uptake of soil nutrients is held or retained in the forest biomass as new wood. Most of it helps to make new leaves. Even although the total is small, twice as much potassium and phosphorus is retained in the trunks and branches of deciduous forest as in coniferous forest.

Note that although the annual retention of elements in the wood of trees is fairly small, the *cumulative* effect of stored nutrients on trunk-growth is considerable, as shown by the large size of forest ecosystems.

(iii) *From biomass to litter*. It is clear from the above analysis that a large fraction of the nutrients taken up each year by the two forest systems is returned to the soil each year in the form of leaf-litter. For example, five-sevenths of the potassium and five-sixths of the phosphorus are returned in this way by deciduous woodland.

(b)   Tropical rain forest

Nutrient circulation is very rapid in the tropical rain forest as a result of very fast processes of growth and decay under constantly hot, wet, tropical conditions. The annual return of nutrients to the soil, for instance, is some 3–4 times that of temperate deciduous forests, where a short growing season restricts both growth and decay. Nutrient cycling is not only rapid but also highly efficient. Tropical forests have a shallow (40–50 cm), but very dense, surface root-system which efficiently absorbs nutrients released from surface litter-decay and prevents them from being leached downwards. *Mycorrhizae* (i.e. fungal growths), which are associated with the roots of many tropical plants, are also thought to improve the absorption of both water and nutrients, especially from soils poor in nutrients.

ASSIGNMENTS
1.   *Refer to Figure 6.1.*
     (a)   *Compare the relative distribution of nutrient reserves in boreal coniferous, cool temperate deciduous and tropical rain forest biomes.*
     (b)   *Explain why there are differences among these biomes in the storage of nutrients.*
2.   (a)   *Use Figures 6.1 and 6.2 to justify the claim that nutrient circulation is greater in deciduous hardwood than in coniferous softwood forest.*
     (b)   *What evidence is there that nutrient cycling in tropical rain forest is fast and efficient?*

C.   Human Usage and Biome Response

Introduction

The world's biomes are used by humans in many different ways. The use of an individual biome is influenced by its productivity and by the distribution of its reserves of nutrients. For example, the large biomasses of the boreal forest and tropical rain forest form an important resource of timber today, while the large reserves of soil nutrients of the temperate grassland biome sustains many of the world's most important grain-producing areas.

As well as attracting different human activities, the various biomes also vary in their ability to 'tolerate' these activities. Some ecosystems, such as the tropical rain forest, the semi-arid grasslands and the tundra, are fragile when faced with the impact of present-day human actions. On the other hand, temperate deciduous forest and moist temperate grassland ecosystems appear to be much more hardy under human occupation.

1. Tundra

The tundras, which have low plant productivity and slow rates of nutrient circulation (see Figure 6.3) under very cold, wet conditions, have been little used by humans. For many millenia, the Arctic tundras have been the homeland of peoples such as the Inuit (Eskimos) and Sami (Lapps) whose way of life is based upon hunting, fishing and reindeer-herding. Because of the vastness of the tundra, their activities have not disturbed the ecosystem to any great extent.

In recent years, however, tundra habitats have been subjected to increasing disturbance. In areas such as Alaska and north Canada, the search for and exploitation of fuel and metal deposits have meant the construction of mines, pipelines, roads and airstrips, all of which badly affect the natural environment. In Alpine-type tundras (e.g. Switzerland, Scotland), serious disturbance has been caused by increasing pressure from tourists (e.g. walkers; skiers and ski-lifts).

The tundra ecosystem is fragile because there are so few living plant and animal species (some of which may become extinct) and because the rate of growth of vegetation is so slow. The soils of the tundra are thin, peaty and immature, and are also slow to form. Therefore, when vegetation and soil are damaged, they have very poor rates of recovery. In the Arctic tundra, for instance, tracked vehicles easily remove the plant cover of moss, lichen and grass. The soil is therefore exposed to increased insolation (heat), so that the permafrost thaws to greater depths. This results in soil subsidence and erosion and the making of hummocky, 'thermokarst' landscapes. Once vegetation and soils are destroyed in this way, they take a long time to recover.

2. Arid and semi-arid deserts

(a) The promise of irrigation

Like the tundra, the semi-arid and arid desert biomes are largely unproductive, with low rates of nutrient circulation (see Figure 6.4). Most of the extremely arid areas have been barely touched by human activities, which instead have been concentrated in patchy, more favourable locations where irrigation water can be applied. Such areas include: (i) along

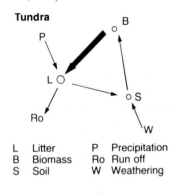

Figure 6.3 Nutrient cycling in the tundra, illustrating a low rate of both circulation and storage. Inputs from weathering and outputs from run off are also negligible. (Source: Gersmehl, 1976)

L Litter	P Precipitation
B Biomass	Ro Run off
S Soil	W Weathering

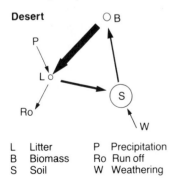

Figure 6.4 Nutrient cycling in arid deserts, showing slow rates of circulation but a large potential reserve of plant nutrients in the soil (Source: Gersmehl, 1976)

L Litter	P Precipitation
B Biomass	Ro Run off
S Soil	W Weathering

142

rivers whose waters come from more humid areas (e.g. the Nile); (ii) in oases where the water table, being close to the surface, increases plant productivity (e.g. around the Ahaggar massif in the central Sahara); (iii) along the shores of coastal deserts (e.g. Peru).

It is clear from the above examples that the desert, if water can be brought to it, can become biologically very productive. The high productivity of some irrigated desert areas is based on high amounts of insolation, together with the large, hidden store of nutrients in desert soils (see Figure 6.4).

Making the desert fertile by bringing in irrigation water can be self-defeating, however. Irrigation is a demanding technology. Many schemes have been poorly designed and managed and are now becoming unproductive. One of the main reasons for the failure of irrigation schemes has been their inability to cope with *salinisation*.

(b) Salinisation

(i) *Process and effects*. The Colorado River in south-west USA is one of the most developed watercourses in the world. A complex system of reservoirs and nine major dams helps to irrigate over one million hectares. River water, unlike rain water, always contains a certain amount of salt. When water from the Colorado River is used for irrigation, most of its salt ends up on the land (see Figure 6.5). As the irrigation water evaporates from the surface, it leaves behind a crust of

Figure 6.5   Process of salinisation

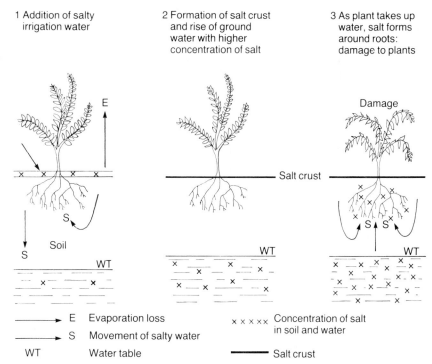

143

salt and the rest sinks down to the groundwater, slowly making it more salty. As plants take up water, salt also concentrates around their roots and the soil becomes salty enough to damage them. (See Plate 5.12 on page 129.)

Today 10% of the world's irrigated lands are seriously affected by salinisation. Vast areas of salinised land can be found in the Middle East, the USSR, India and Pakistan, the western United States, Egypt, China and Australia. It has been estimated that more food is now lost as a result of irrigation schemes becoming unproductive by salinisation than is grown as a result of costly new irrigation works. In many developing countries, salinisation leads to crop failure and famine and to the land being abandoned.

(ii) *How to stop salinisation*. To cure the build-up of salt, farmers can flood the land with a lot of water which will wash away the salt (see Figure 6.6). If the land is not properly drained, this action only makes the salty groundwater rise. When it comes within reach of the roots of the plants, they will die. However, if drains are laid, they will lower the groundwater level. The salt can then be flushed out of the soil continuously and the plants will flourish.

Usually the drainage water has to be pumped out of the ground and into open drains. Farmers in the areas irrigated by water from the lower Colorado attempted to solve their salt problem by pumping drainage waters back into the Colorado River, just above the Mexican border. The Mexicans complained bitterly about this to the US government because the river water entering their territory had become too salty to use for

Figure 6.6 Counteracting salinisation: applying large quantities of water under poor and good drainage conditions

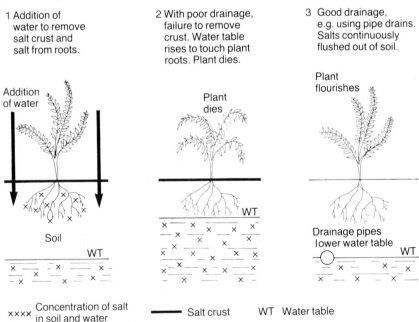

1 Addition of water to remove salt crust and salt from roots.

2 With poor drainage, failure to remove crust. Water table rises to touch plant roots. Plant dies.

3 Good drainage, e.g. using pipe drains. Salts continuously flushed out of soil.

×××× Concentration of salt in soil and water   —— Salt crust   WT Water table

irrigation. Now a $200 million desalinisation plant has been built by the US government to remove salt from irrigation drainage waters emptying into the Colorado.

It is clear from the above that salinisation can be cured, but it costs a lot of money to put in drains, etc. Drip- or trickle-irrigation, where water is piped underground directly to plant roots, uses little water and can be extremely efficient. But this too is expensive.

There is little doubt that many more irrigated areas will become affected by salinisation over the next few decades. They will become unproductive and will ultimately revert to desert and scrubland.

3. Forest biomes

Because the major forest biomes differ very much in their productivity and patterns of nutrient circulation, they have been exploited for a variety of purposes. In addition, the different forest biomes have responded in contrasting ways to human intervention: some appear hardy and are able to withstand whatever use to which they are put; others appear fragile, and are severely damaged by modern activities.

(a) Boreal forest

Forests are normally found in areas where there is enough rainfall for sustained agricultural production. In the case of the boreal forest biome, the low reserves of soil nutrients in its predominant soil (the podsol), together with extreme cold and a short growing season, have prevented large-scale or long-term agricultural development. The region has been used, however, for its resources of animals, minerals, power and timber.

The native inhabitants of the boreal regions were scattered tribes or families, living by hunting caribou, seals, moose or fish. Large-scale exploitation of the wealth of this forest started in the sixteenth and seventeenth centuries when trappers were attracted by the many furry animals, such as beaver, sable, mink and fox. Widespread removal of industrial materials (e.g. minerals, timber) from the boreal forest region began in the nineteenth century. The ancient crystalline rocks which lie below much of the boreal forest are rich in metal ores; younger, sedimentary rocks contain deposits of coal, oil and gas. Glaciated areas, with huge fresh-water systems of lakes and rivers which are fed by plentiful snow-melt, form the basis of large-scale, hydro-electric power schemes. All of these conditions, combined with the greatest resource of the biome – the seemingly endless reserves of softwood timber – have resulted in a great development of mining, hydro-electric, lumber and paper industries.

(b) Temperate deciduous forest

In contrast to the boreal coniferous forest, much of the deciduous forest biome has been cleared for agricultural use because the brown soils of the temperate forest are fertile. They contain large reserves of nutrients (see Figure 6.1b). When the forest is cleared, the soils are able to support long-term arable and livestock production. Also, climatic conditions are more suitable for agriculture (see Figure 5.7 on page 112). In the temperate deciduous forest, a plentiful supply of well-distributed rainfall, together with sufficiently high summer temperatures, allows high crop-growth and productivity.

(c) Tropical rain forest

Tropical rain forest biomes are of much less use for agriculture. There are three reasons for this.

(i) *Removal of biomass*. Unlike the forest systems of temperate areas, the bulk of the reserve of nutrients of the tropical rain forest is held in the living biomass rather than in the soil or litter (see Figure 6.1c). The removal of the wood by forest clearance means a large and instant loss of the system's nutrient reserves (see Figure 6.7).

(ii) *Erosion and leaching*. Removing the forest and replacing it with grassland or cropland exposes the soil to the ravages of the tropical climate. For instance, under conditions of high temperature and humidity, the remaining litter rapidly decomposes because of increased bacterial activity (see Figure 6.7). Without the protection of the multi-layered forest canopy, the land is open to the effects of heavy tropical rainfall which creates larger amounts of surface run off and, eventually, a loss of nutrients by soil erosion. Reserves of nutrients are also removed by increased leaching: high rainfall, together with lower rates of evapotranspiration in the absence of forest, encourages the downward movement of water through the soil and thus the translocation of soil nutrients. This leaching leads to the removal of silica and bases and helps to accumulate iron and aluminium at the surface. Eventually a hard surface crust of laterite may form.

(iii) *Formation of laterite*. Without a forest canopy, which previously provided shade, the degraded parent material can be easily dried out,

Figure 6.7   Interference in the nutrient cycle of tropical rain forest by forest clearance and cultivation

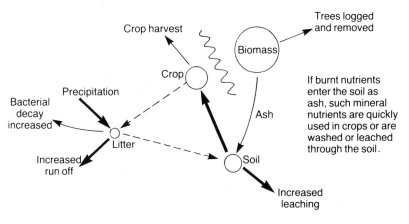

146

producing hardened, brick-like, laterite crusts. *Laterisation*, or the formation of lateritic plinthite, is becoming a serious problem in the moist tropics and in the savannas, since it causes long-term and often irreversible change in the ecosystem. Lateritic crusts and pavements are not easily replanted with forest, nor cultivated.

4.   Grassland biomes

(a)   A resilient ecosystem

The semi-natural grasslands of the world are especially adapted to grazing by large herbivores. The tropical savannas in particular are grazed by many large, hoofed animals, including wildebeest, zebra, antelope, and introduced livestock, especially cattle. However, cattle grazing, or ranching, is more typical of the temperate grassland (e.g. prairies, pampas). Grasses are well adapted to grazing, cutting, trampling and light burning: unlike many shrubs and herbs, whose leaves grow from their tips, the growth tissue of grass is located at the *base* of the shoot, close to the soil surface. Therefore, leaf formation and growth can continue after, and is often encouraged by, grazing and cutting.

The grassland biome is also cultivated wherever possible. In this respect, the temperate grasslands seem more suitable and resilient than the tropical savannas. We can find the reason for this by comparing the reserves of soil nutrients of the two grassland biomes. Figure 6.8 shows that the chernozem soil of the temperate grassland is far richer in nutrients than the latosol of the savanna. Because the temperate grassland soil is so fertile, the more moist grasslands have developed this century from being open pastures for animals into large-scale, grain-producing areas, such as the Soviet steppes and the North American prairies.

Figure 6.8 Nutrient cycling in the temperate steppe grasslands and tropical savanna grasslands (Source: Gersmehl, 1976)

(a) Steppe

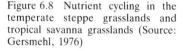

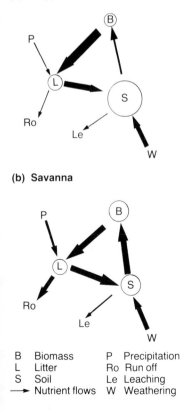

(b) Savanna

B Biomass	P Precipitation
L Litter	Ro Run off
S Soil	Le Leaching
→ Nutrient flows	W Weathering

(b)   A fragile ecosystem

The grassland biome is less resilient, under both grazing and cultivation, in the drier areas with more marginal rainfall. In Africa, the traditional users of the semi-arid grasslands are the nomadic pastoralists who herd their cattle across the desert to find pasture where rain has fallen and who cultivate only small areas. This is the most efficient means of using the sparse productivity of the area, but it depends on keeping down the numbers of cattle and people to levels that can be sustained. Now the delicate balance has been disturbed: in recent years, increasing population, over-grazing, over-cultivation and widespread wood-cutting have reduced these marginally usable grasslands to desert.

(c)   Desertification

(i) *The process of desertification.* When the quality of vegetation and soils is reduced in semi-arid and sub-humid regions, more desert-like surfaces result. This is what is meant by 'desertification': the making of deserts.

147

Desertification begins with a decline in the vegetation cover, either because of climatic drought or because of its removal by human actions. Less vegetation cover produces less organic matter to maintain soil structure and to help the soil to hold water and nutrients. The reduced vegetation cover no longer protects the soil, which dries out and the fine organic matter and mineral particles are either blown or washed away. When the rain does come, the wet soil is baked by the sun to form a crust on the surface, which prevents water from sinking in. Compaction of soil, loss of nutrients, removal of soil cover, salinisation and a loss of surface water and ground-water may all be part of the process.

Desertification is a global problem, with two out of every three nations, both rich and poor, being affected. In south-west USA, 40 million hectares are seriously damaged; in India one third of the arable land is threatened with the total loss of topsoil. The African continent has suffered most from desertification (see Figure 6.9). It was first brought to the world's attention by the drought and famine in the Sahel (at the southern edge of the Saharan desert) in the early 1970s. Its consequences have been shown by the continuing famines of the 1980s in Burkina Faso (formerly Upper Volta), Sudan, Somalia, Ethiopia. Seven million km^2, an area twice the size of India, is at risk in Africa south of the Sahara. During the period 1950–1975, drought and poor farming practices in the Sudan pushed the desert south by 100 km.

As shown in Figure 6.10 pockets of desert sometimes appear next to, and advance along a wide front from, the arid and semi-arid zones. The

Figure 6.9   Extent of global desertification (Source: Barke and O'Hare, 1984)

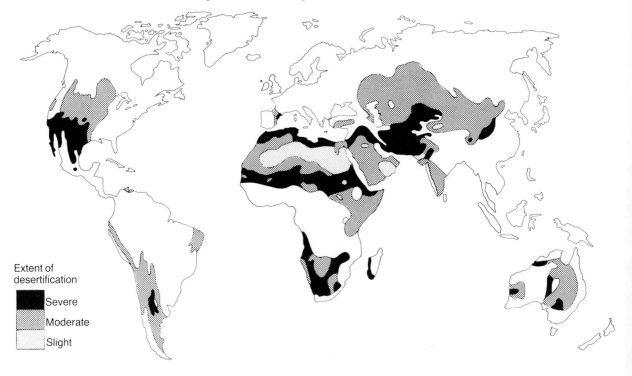

Extent of
desertification

Severe

Moderate

Slight

Figure 6.10   Expanding desert in Burkina Faso, sub-Saharan Africa

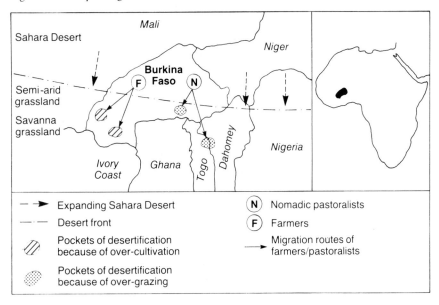

process may also occur far away from the arid areas. Breaking out in small pockets, it can spread until it lays waste large areas, as in the moist grasslands of southern Burkina Faso.

(ii) *What causes desertification?* Desertification is made worse by *climatic drought*, especially in areas with already low supplies of moisture. There is little doubt that lower-than-average rainfall in Burkina Faso during the last 20 years has added to the desertification problem (see Figure 6.11). However, drought is not the chief cause of desertification. The major causes, particularly in Africa, are the rapidly increasing population and the lack of investment in agriculture and forestry. There are four kinds of mismanagement.

Figure 6.11   Trends in the average rainfall of Burkina Faso (1915–1985), showing the recent decline in amount. Annual rainfall values are expressed as a statistical variation (standard deviation) from the overall mean represented by the line 0–0 (Source: Hartzell, 1986)

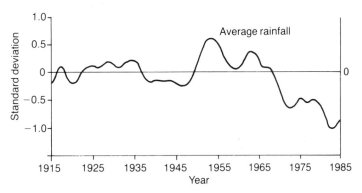

1. *Forest clearance* is causing desertification in most African and Asian countries. Trees are essential for protecting the soil from sun, wind and rain. Their deep roots provide channels for water to percolate down, so that it can be held in the soil. The main reasons for rapid forest removal are the use of forest areas for farmland and the demand for fuel in urban areas. For instance, almost all the trees within 40 km of Ouagadougou, the capital of Burkina Faso, have been felled.

2. *Over-cultivation.* In sub-Saharan Africa, neither governments nor aid agencies (e.g. the World Bank) have invested much money in efforts to improve poor farming methods. As a result, the only way to feed extra mouths has been to expand the area under cultivation. As the demand for land increases, the fallow or resting periods, which should be anything between 8 and 15 years, have been reduced more and more. Shorter fallow periods mean that there is not enough time to allow the soil to recover lost nutrients during the re-growth of the fallow vegetation. As yields decline, farmers have been forced to bring into production other marginal areas, e.g. areas with steep slopes and thin soils. These areas are easily turned into desert.

3. *Over-grazing.* In many countries, livestock numbers have increased while the amount of grazing land available has declined because of the demand for cropland. Large numbers of livestock quickly outreach the ability of the land to support them; over-grazing, with stripped or flattened vegetation, is the result (see Plate 5.11 on page 128). In sub-Saharan Africa, 20 million people are nomadic pastoralists, getting much of their livelihoods from cattle, goats and camels fed on naturally available forage. Now there are too many cattle and goats, and the balance between humans and the land has been broken. Many of the nomadic pastoralists, displaced by desertification in the north, are migrating south to find new pastures in the more moist, savanna grasslands. These 'ecological' refugees are now upsetting the balance between humans and the land in the south: as shown in Figure 6.10, their extra numbers are speeding up desertification in the sub-humid savanna grasslands.

4. *Salinisation*, or the production of salt deserts, is also a cause of desertification in its widest sense. Salinisation, which results from the poor management and design of irrigation projects, is outlined above in section C.2.

(iii) *Restoration measures in the USA.* Desertification has occurred and is occurring today in North America. In the prairies in the 1930s (and in the Russian steppes in the early 1950s), the removal of the natural vegetation on the drier fringes, followed by over-cultivation and over-grazing, caused widespread loss of soil by wind- and water-erosion. Serious soil erosion, especially by wind (i.e. the 'dust-bowl' effect), resulted in the withdrawal of wheat-farming from the drier, more drought-ridden areas. Measures leading to the restoration and conservation of the prairies were begun in the 1930s and continue to the present day. Under government guidance, grazing on re-seeded land was established. The planting of wind-breaks and shelter-belts, together with contour ploughing, have proved important for soil conservation. On grassland, various techniques of range management are now used, such as the prevention of over-grazing and the control of undesirable, 'invading' plants. Sadly, no such measures are widely practised, as yet, in the savannas and semi-arid tropical grasslands.

ASSIGNMENTS

1. (a) *Using Figures 6.1 and 6.2, describe and explain the contrasting human usage of temperate deciduous and boreal coniferous forest biomes.*

   (b) *Refer to Figures 6.1 and 6.7. Give reasons why the agricultural potential of land in the moist tropics which is cleared of forests differs from that in cool, moist, temperate environments.*

2. (a) *Define the term 'desertification'.*

   (b) *Using Figures 6.9, 6.10 and 6.11, describe the causes and distribution of the desertification problem.*

Key Ideas

A. *Plant Productivity of the World's Biomes*

1. Plant productivity or growth rate varies between the major global land ecosystems.
2. Generally, forests are more productive than grasslands, which in turn are more productive than tundra and arid desert biomes.
3. The growth rate of vegetation is determined by the length of the growing season, which is influenced by environmental factors such as the availability of moisture, energy and nutrients.
4. It is also determined by the physical size and cover of the vegetation, since this acts as a trap for solar energy.

B. *Nutrient Cycling in Forest Biomes*

1. Most of the reserve of nutrients of the boreal forest is contained in the litter compartment, while the bulk of the nutrients in the tropical rain forest are held by the living forest biomass.
2. There is a much more even distribution of nutrients, among soil, biomass and litter, in the temperate deciduous forest biome.
3. Nutrient cycling is greater in the temperate deciduous biome than in the boreal coniferous forest biome.
4. The tropical rain forest has an exceedingly fast and very efficient nutrient cycle.
5. Within forest ecosystems, only small amounts of nutrients are stored in the annual production of new wood but the cumulative effect on trunk-growth is considerable.

C. *Human Usage and Biome Response*

1. The human usage of individual biomes is influenced by their levels of productivity and by their distribution of nutrient reserves.
2. Low plant productivity and slow rates of nutrient cycling limit the human use of tundras and arid deserts.

3.   Arid and semi-arid deserts can be damaged by poorly managed irrigation projects as a result of salinisation or the concentration of salts in the soil surface layers.

4.   The main organic resource of the boreal forest lies in its biomass, i.e. in its vast reserves of softwood timber.

5.   Temperate deciduous forest and temperate grassland biomes have been valued more for their soil or agricultural resources.

6.   The tropical rain forest is a fragile ecosystem with limited agricultural potential after trees have been claered.

7.   Many grassland areas are well adapted to grazing by large herbivores, including livestock.

8.   Climatic drought together with human misuse of the land have reduced marginally usable, semi-arid grassland areas to desert.

9.   Desertification, or the creation of desert-like surfaces, is made worse by over-grazing, over-cultivation, widespread wood-cutting and poor irrigation.

Additional Activities

1.   (a)   Using Figure 5.3 on page 109, draw an outline map of the world showing the distribution of the major world biomes.

      (b)   Using the data in Table 6.1, construct a chloropleth map of world biome productivity (growth rate). A convenient scale is: <0.1 kg/m²/yr;   0.1–<0.2;   0.2–<0.6;   0.6–<1.0;   1.0–<1.4; 1.4–<1.8; >1.8 kg/m²/yr.

      (c)   Refer to the climatic diagrams associated with each major biome in Chapter 5. Calculate the vegetation productivity associated with each climatic zone by using the formula

$$I = \frac{T_{m}.P}{120T_{r}}$$

where I = index of plant productivity (estimated)
  T_{m} = average temperature of the warmest month (°C)
  T_{r} = annual temperature range between warmest and coldest months (°C)
  P = annual precipitation (cm)

      (d)   Examine the relationship between the measured productivity values in Table 6.1 and your own calculations, noting any significant differences.

      (e)   What factors may account for variation in the two sets of data?

2.   (a)   What is meant by a fragile biome?

      (b)   Define laterisation, desertification and salinisation.

      (c)   Compare and contrast these processes in terms of: (i) the areas where they are most likely to occur; (ii) how they operate; (iii) the environmental factors involved in their formation.

      (d)   Give reasons why they are difficult to reverse or remedy.

7 Whither Ecosystems? The Human Impact

This chapter draws together a wide range of examples involving human intervention in ecosystems. Section A illustrates how humans can remove ecosystems, by describing the loss of the traditional British countryside as a result of modern farming and forestry. The next section describes how the global nutrient cycle is affected by acid rain. The origin, dispersal and deposition of acid rain over Europe are examined, together with the environmental consequences and the possibilities of control. Section C looks at global soil erosion, with examples from Nepal and Great Britain.

A. Britain's Lost Landscapes

1. Agents and scale of destruction

There is an immense variety of habitats in the UK because of a very wide distribution of rock types, landforms and climate. As a result, in a relatively small area, Britain supports semi-natural landscapes as varied as Arctic-Alpine heath, upland moorland and bog, ancient woodland, chalk downland and lowland meadow and marsh. Yet all of these habitats, and virtually every species that depends on them, are today under threat. Studies have shown that nearly all of the habitats important for wildlife have been reduced considerably in area in the post-war period. In the past 40 years, the British countryside has lost 95% of its flower-rich hay meadows and 80% of its pasture on chalk downs; 40–50% of its ancient, semi-natural woodland has been felled and replaced by foreign conifer plantations; 60% of the heathlands and more than half of the marshes have gone. More than 200 000 km of hedgerow – enough to encircle the equator five times – have been removed and levelled to make way for bigger fields (see Figure 7.1).

The chief agent responsible for destroying the traditional British countryside and its wildlife is modern intensive farming (see Chapter 1, Section F). Modern forestry, with the increasing number of conifer plantations, and the water industry, with the building of large reservoirs, are important secondary causes of the destruction of traditional landscape. Urban growth and the increased impact of the recreational activities of tourists have generally proved less of a threat to the British countryside than modern farming and forestry. Figure 7.2 shows the main areas of countryside (and wildlife) that will probably be threatened by modern intensive agriculture and forestry in the next ten to twenty years. Blank areas are those where no significant tracts of semi-natural habitats survive.

Figure 7.1 Percentage loss of species-rich, semi-natural habitats in the UK since 1945

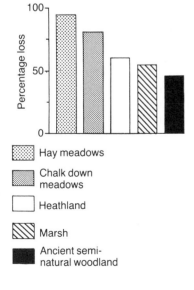

Hay meadows

Chalk down meadows

Heathland

Marsh

Ancient semi-natural woodland

Figure 7.2 The major areas of countryside in Britain that will probably be under threat within the next ten or twenty years. (Source: Pye-Smith and Rose, 1984)

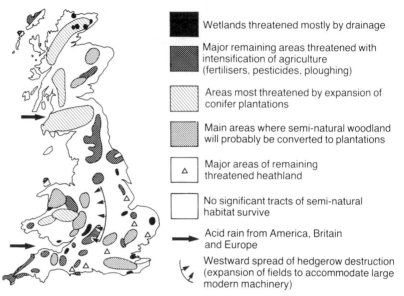

Successive British governments, in their drive to increase productivity and self-sufficiency in certain basic foodstuffs (e.g. wheat, milk, meat), have constantly supported the farming industry. Farmers are given guaranteed prices for their crops, and are provided with a wide range of grants, loans and tax-free concessions for expanding the culti-vated area, improving land-drainage, and removing hedgerows. In 1982 the government of the UK spent over £400 million on agricultural improvement and intensification. During the same year a mere £4 million was given to the Nature Conservancy Council, the principal conservation body in the UK, to protect wildlife and its habitats.

2. Destruction of habitats

(a) Old grassland and meadow

Old, unimproved grassland and meadowland are species-rich habitats (especially for butterflies) which have suffered considerable losses in recent years. In the new county of Avon (near Bristol), 50% of the unimproved meadows was destroyed between 1970 and 1980; 69% of Devon's grassland was lost between 1952 and 1972. As Figure 7.3 demon-strates, Dorset has lost almost 75% of its species-rich chalk downland between 1811 and 1972, most of it vanishing under the plough in the last 50 years. (See Plates 1.2 and 1.3 on page 21.) Starvealla Farm, on Oxfordshire's Wendlebury meadows, contains one of the most important medieval pasture-lands in Britain. It is recognised as of international importance for conservation and is designated as a top-class Site of Special Scientific Interest (SSSI) by the Nature Conservancy Council. But, in August 1980, some 12 hectares were ploughed up by a local

Figure 7.3 Vanishing habitats of the Dorset heath and downs, 1811–1972 (Source: Pye-Smith and Rose, 1984)

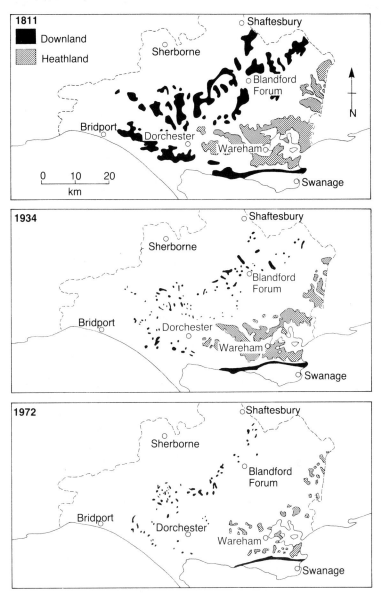

farmer. A rich, varied flora and fauna was converted to a uniform cereal crop (monoculture) the following year.

(b) Moorlands and heaths

Many of Britain's shrubby heaths (heather-dominated vegetation) are disappearing fast. The upland heather moorlands are rapidly being planted with trees: a staggering 90% has been destroyed in Dumfries and

Galloway since the 1930s; in the last 40 years, more than 20% of Dartmoor has disappeared. In the same period, over 20% of Exmoor has been claimed for improved pasture and cultivation.

In the lowlands, acid heaths are being reclaimed for agriculture. Figure 7.3 shows that 75% of the heathland of Dorset has also vanished over the last 50 years. Much of the area lost has been ploughed up, limed, freely fertilised and generally 'improved' for cereal production.

3.   Reduction in distribution of species

When local habitats, such as meadows and moorlands or ancient woodlands and lowland bogs, are removed, the plant and animal species associated with them also die. Loss of habitats on a national scale reduces the population and area of distribution of individual species. If the overall decline is drastic, the species may be lost completely; it will become *extinct* or be on the verge of extinction, i.e. it will become an *endangered* species.

(a)   Extinction of species

A notable example of a species becoming extinct because of loss of habitat is the Large Blue Butterfly. This beautiful butterfly, at one time found in Northamptonshire, south-west England and the Cotswolds, finally became extinct in Britain in 1979. For it to survive, it needed very short grass on warm slopes, such as exists on the chalk downlands. Changes in farming (ploughing, fertilisation, herbicides) and new forestry plantations destroyed many favoured sites. Others sites disappeared beneath encroaching scrub after the rabbit, which had always maintained a close-cropped turf, became a victim of myxomatosis, a disease deliberately spread by humans.

(b)   Endangered species

The Monkey Orchid is now an endangered species. Ploughing, building development and plant-collecting have combined to put the handsome Monkey Orchid on the danger list. Now known in only three places, two in Kent and one in Oxfordshire, the orchid is reduced to fewer than 50 individual plants.

There has been a great reduction since 1950 in the range of another plant, the Marsh clubmoss (see Figure 7.4). This species is dependent on lowland wet heaths, many of which have been drained, fertilised and ploughed up in recent years. The present distribution of the plant is now so reduced and scattered that it must be considered as endangered.

Many other wild species of plants, mammals, insects and birds have become endangered in recent years. A full list of these can be found in the *Red Data Book* published by the International Union for Conservation and Natural Resources (IUCNR), Switzerland.

Figure 7.4   Reduction in the range of the Marsh clubmoss in relation to the loss of lowland wet heath, its habitat. Open triangles denote its presence, in 10 km-squares of the National Grid, before 1950 but now no longer present. Filled circles record its location before 1950 or since 1950 and still present today. (Source: Nature Conservancy Council, 1981)

△ Recorded before 1950
  but no longer present

● Recorded before 1950
  or since 1950 and still present

Marsh clubmoss

4. Conservation measures

Against the above losses of habitat and wildlife, there have been some notable gains over the last 40 years. In England and Wales ten National Parks and a much bigger 'second division' of Areas of Outstanding National Beauty (AONBs) have been established by the Countryside Commission. This organisation was set up to preserve the beauty of the countryside. In the same period the Nature Conservancy Council (NCC), despite working on a shoestring budget, has set up throughout Great Britain a network of National Nature Reserves (NNRs) and has listed some 4000 Sites of Special Scientific Interest (SSSIs) – some of which are also Nature Reserves – to protect our last remaining habitats.

For instance, Gibraltar Point, Lincolnshire, is a well-protected Nature Reserve (see page 71). Here 433 hectares of dune and salt marsh have been in the care of the Lincolnshire and South Humberside Trust for Nature Conservation since 1949. Thousands of wildfowl visit the Reserve; the vulnerable little tern nests safely. Rare plants include the 'sea blight' which is found on the young salt marshes.

However, designating such areas is no guarantee of protection. About 10–20% of the SSSIs, for instance, have already been destroyed or seriously damaged. Because of this, the NCC is presently fighting for tougher legislation and more effective protection against development. It has urged that the farming community should not reclaim the last remnants of good wildlife habitats (e.g. small woods, old hedges and ponds) that still exist on their farms. It has recommended that farmers should be encouraged to obtain their extra production and profits by farming the existing cropped land more intensively.

ASSIGNMENTS
1. (a) Using Figure 7.3 and the text, examine the loss of unimproved grassland and heathland in the UK.
    (b) List four other semi-natural habitats which have suffered a major decline in recent years.
    (c) Using Figure 7.2, describe the role of modern agriculture and forestry in the decline of habitats.
2. (a) How may the distribution and survival of a species be affected by the loss of its habitat? Give examples, referring to Figure 7.4.
    (b) Can you list any other wildlife species, not mentioned in the text, which are threatened by modern 'development'?
3. Can you think of any reasons why: (i) habitats and their wildlife should be conserved; (ii) cities pose less threat to the countryside and its wildlife than modern agriculture; (iii) large gardens and roadside verges are important wildlife habitats?

B. Acid Rain over Europe

1. What is acid rain?

'Acid rain' is a general term for all kinds of air-pollution. Not all of the pollutants which form acid rain are acidic (e.g. ozone) nor do they always

Figure 7.5 The production, transport and effects of 'acid rain' on the environment (Source: McCormick, 1985)

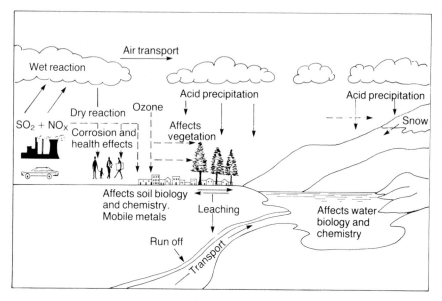

Figure 7.6 Estimated emissions of sulphur dioxide in Europe, by country, in 1980 (Source: Bunyard, 1986)

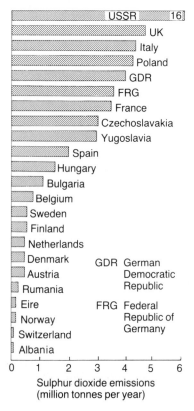

Sulphur dioxide emissions (million tonnes per year)

fall as rain. Sometimes 'acid rain' is deposited at the earth's surface by snowfall and snow-melt, or as a deadly acid mist. Sometimes the pollution settles directly on trees, buildings and the ground as a 'dry' deposit, i.e. without going into solution (see Figure 7.5). Although there is a complex cocktail of pollutants involved in acid rain, the main ingredients are sulphur dioxide and nitrogen oxides, which are given off or emitted mainly by coal-burning power stations and motor cars. These gasses dissolve in cloud and rain, to form weak sulphuric and nitric acid. Nitrogen oxide mixes with other chemicals under sunlight to form ozone, another gas which is a particularly unpleasant pollutant. In the UK it is estimated that about 67% of the acid rain comes from sulphur dioxide, while most of the other 33% comes from nitric acid. An examination of the emission, transport and deposition of sulphur dioxide thus provides important clues to the problem of acid rain.

2. The role of sulphur in acid rain

(a) Emissions

As a result of the industrial revolution, the quantities of sulphur dioxide which have been released into the atmosphere by most countries are very large. In Europe, the USSR probably heads the list with about 16 million tonnes each year, followed by Britain with some 4.7 million tonnes. Most of the other European countries have annual emissions in the range of 2.0–4.5 million tonnes (see Figure 7.6). However, emission levels of sulphur dioxide in Europe are tending to level off as a result of cut-backs in industry and/or pollution control. For instance, in the UK, emissions of sulphur dioxide reached a yearly maximum of about 6 million tonnes in the mid-1960s and have since started to fall because of a decline in heavy industry.

While the amounts of sulphur dioxide produced over Europe are stabilising, it is important to realise that those of the other acid pollutants are not. Emissions of nitrogen oxides, for example, have continued to rise. The most likely sources of these are exhaust emissions from road vehicles and industry, and ammonium compounds released during the use of nitrogen-rich fertilisers. These facts suggest that we should not be complacent about the likelihood of automatic reductions in acid-rain fallout.

(b) *Transport and deposition*

One of the essential features of atmospheric pollution is that it is often concentrated, and has effects, in areas beyond the source of its production. Air-pollution by sulphur dioxide, especially when emitted high into the atmosphere from tall chimney stacks, is easily carried by winds to neighbouring regions and deposited there.

Figure 7.7, which shows the amount of sulphur deposited in Europe in 1980, also highlights the broad patterns of export and import of sulphur. Although about 20% of the sulphur deposited in Britain comes from neighbouring countries, Britain exports about 70% of its sulphur output to other countries, mostly to central Europe and Scandinavia. While the Scandinavian countries of Norway, Sweden and Denmark are relatively small sources of sulphur emission (Figure 7.5), they are major importers from other countries: 92%, 82% and 64% respectively of the sulphur deposited come from areas outside their borders. A third group of countries, including West Germany, Poland and Czechoslovakia, can be identified as major producers *and* importers of sulphur. As a result, they receive the highest total fallout of sulphur, with 1.4–1.6 million tonnes annually.

3. Effect on ecosystems

Rain and snow falling over many parts of the northern hemisphere (including Europe) are being made more and more acid by atmospheric pollution. Rainwater is normally slightly acid, with a pH of around 5.6. By the early 1970s, mean pH values had fallen a full unit to about 4.6, suggesting a ten-fold increase in acidity (see page 18). Such acid rain has many effects on individual ecosystems: it causes, as shown in Figure 7.5, direct damage to building materials through corrosion, and has a wide range of direct and indirect effects on crops, forests and soil as well as on rivers and lakes.

(a) *Acidification of soils and lakes*

As shown in Figure 7.5, much of the rain which falls at the surface passes over vegetation and through soils before reaching rivers and lakes. In draining through soil, acidified rainwater leaches out the soil bases,

Figure 7.7 Source and amount of sulphur deposited in Europe in 1980 (Source: EEC, 1983)

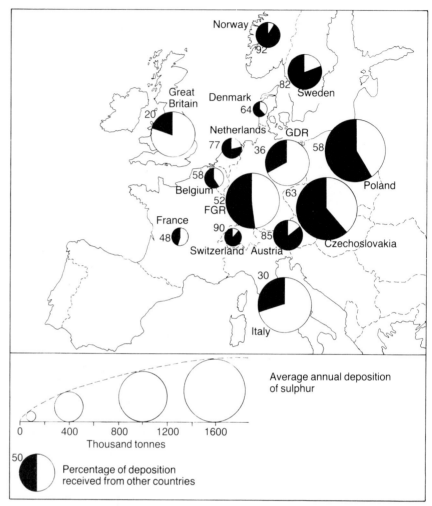

Figure 7.8 The change of acidity (pH) at different depths in podzols and grey-brown forest soils in southern Sweden, 1949–1984 (Source: Bunyard, 1986)

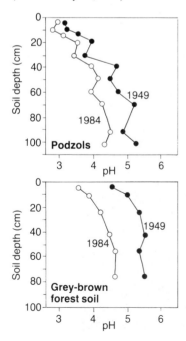

replacing them with hydrogen. It is not surprising, therefore, that many soils in Sweden, including both podzols (below coniferous forest) and brown soils (below deciduous forest), show a considerable decrease of pH compared with measurements made 40 years ago. Figure 7.8 shows that the average pH decrease is 0.7–0.9 units throughout the whole profile of the brown forest soils and the B-horizon of the podzols. This decline shows that about one-half of the base cations, including calcium, magnesium and potassium, has been lost from the soil and replaced by hydrogen.

Many lakes and streams in Sweden, West Germany and even Scotland have been made more and more acid in recent years, with pHs normally about 6.0 dropping below 5.0 (see Figure 7.9). As well as being very acid, many of these lakes have accumulated large concentrations of aluminium which has been washed into them by acid leaching from surrounding soils.

Figure 7.9 Reconstructed acidity of Round Loch of Glenhead, a small lake in south-west Scotland (Source: Bunyard, 1986)

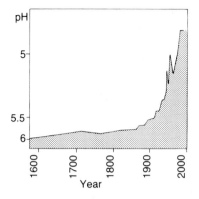

High levels of aluminium are very poisonous in aquatic systems and have been blamed for the massive number of fish killed in these lakes. Salmon and trout have been the species most affected, being sensitive to both acidity and aluminium. About 18 000 Swedish lakes have been poisoned by acid rain; 9000 have lost some fish, and 4000 are virtually fishless. Another 20 000 are at risk.

(b) The death of the trees

Since the late 1970s, trees in ever-increasing numbers have been dying across Europe. One by one, all species are being affected: spruce, pine, fir, beech, oak and ash. If the pace of death continues, large tracts of once-forested land will be left bare of trees.

Factors such as climatic stress (drought, severe winters), disease organisms and the natural acidity produced by the trees themselves (especially conifers) have been blamed for killing the forests. However, the pattern of tree death and decline appears to correspond closely to the fallout of atmospheric pollution. Most forest damage is occurring in Central Europe, e.g. West Germany, Czechoslovakia and southern Sweden, where particularly high levels of acid rain are deposited (see Plate 7.1). In 1982, 8% of the forested area of West Germany showed signs of tree

Plate 7.1 Forest damage by acid pollution in Czechoslovakia (Source: Jiri Pollacek/Panos Pictures)

damage (*Waldsterben* as the Germans call it). In 1983, the affected area was 34% of the total; by 1984 it was 50%. Some experts believe that 90% of West Germany's forests will be dead by early next century.

4. Pollution control

The rapid increase in environmental damage attributed to acid rain has produced a flurry of legislation to control pollution in individual countries. Germany, for instance, has decided to limit drastically the emissions of nitrogen oxides from car exhausts by making it compulsory for motorists to fit special devices to their car exhausts. There have also been moves towards international pollution-control agreements within the EEC and the United Nations. An important step was the formation of the 'Thirty per cent Club' by about 20 nations in Europe, plus Canada. They are determined that by 1993 there will be a 30% reduction in the 1980 levels of sulphur dioxide pollution. The EEC is also proposing a 60% reduction of sulphur dioxide and a 40% reduction of nitrogen oxides from large industrial and electricity-generating plants by 1995.

Britain has remained steadfastly against any significant pollution control. The British government, alone among European governments, maintains that the case against acid rain has not been proven and that more evidence linking acid rain and environmental damage needs to be collected. Such a stance may not be surprising. Britain, located on the western fringe of Europe, does not receive much pollution from elsewhere. It is, however, the major source of emissions of sulphur dioxide in Western Europe. The government's policy of high-stack pollution 'control', where very tall chimneys are built at coal-fired power stations to disperse pollution by sulphur dioxide high into the atmosphere, is relatively cheap and therefore attractive because the government does not at the moment wish to bear the cost of a 'clean-up' campaign. This policy has long angered the Scandinavians who claim that Britain is a major exporter of sulphur dioxide to the rest of Europe and that the acid rain so produced affects their own fresh waters.

But there is increasing pressure on the UK to fall into line with the rest of Europe. For instance, a report in *The Observer* magazine (19 October, 1986) showed that there is now widespread damage by acid rain to Britain's rivers and lakes, wildlife, trees and buildings. In 1986, the British government took a small step forward when it agreed to remove sulphur emissions from three of its eleven coal-fired power stations.

ASSIGNMENTS
1. *Refer to Figures 7.5, 7.6 and 7.7.*
   (a) *What is acid rain?*
   (b) *Describe the pattern of sulphur dioxide emissions shown in Figure 7.6.*
   (c) *Describe the pattern of sulphur deposition shown in Figure 7.7.*
   (d) *Examine the relationship between the two patterns.*

2. (a) *Using Figure 7.5 as a model, describe the general effects of acid rain on the environment.*

    (b) *Examine the link between the death and decline of forests in Europe and acid rain.*

    (c) *How are governments attempting to reduce damage by acid rain to the environment?*

C. The World's Vanishing Soil

1. Introduction

Soil erosion takes place when soil is removed from an area by wind or water at a rate faster than that at which new soil is formed. Wind blows away the finer, usually more fertile, soil particles (fine peat, fine clay); rainfall, acting as run off, can wash soils down slopes. Erosion by rainfall takes two forms: sheet erosion and scour erosion. Sheet erosion takes place when soil is removed from relatively large areas by thin films or sheets of water moving over the surface. In terms of the total amount of soil lost, this is the most important type of erosion by water. Scour erosion is more confined but more spectacular, and happens when run off and loss of soil is concentrated into small channels (rills) or larger channels (gullies) (see Plate 7.2). Soil particles can also be moved down slopes under the influence of gravity, either slowly by soil creep or more quickly by landslide.

2. The causes of soil erosion

Soil erosion is a natural process. Rates of soil erosion under natural

Plate 7.2   Severe gully erosion in north Morocco. This land has been heavily grazed so that only a short, drought-resistant grass cover is present. When the rains come, they easily scour the unprotected, structurally poor soil into rills and gullies. Some patches of low, woody scrub (garrigue) can be seen in the right far ground. (Photograph: G. O'Hare)

conditions are normally very low because there is usually a protecting vegetation cover. Indeed, rates of soil removal are usually balanced by the formation of new soil. But bad farming practices greatly speed up the process of soil erosion. Over the centuries, farming has increased the natural rate of erosion on average by 250%. As much as 2 cm of precious topsoil can be lost in one year.

All over the world, from the edge of the Sahara to China, from North America to Britain, accelerated soil erosion is taking place, often at alarming rates, as a result of human misuse of the land. As shown in Figure 7.10, humans over-exploit the land and cause soil erosion in various ways, e.g. by farming steep slopes, by over-grazing, by continuous growing of arable crops, and by excessive forest clearance. The overall effect on the soil of such misuse is always the same, however. (1) There is a loss of surface soil humus. This weakens the structure of the soil by loosening the individual soil particles (with their nutrients) and preparing them for removal by wind and rain. (2) By cultivating the land, by over-grazing and by clearing forests, humans remove the protective vegetation cover. This exposes the soil to the full power of wind and rain. (3) As erosion proceeds, the plant remains in the soil are lost, and so the soil is further weakened and thus more prone to erosion. A 'vicious circle' of erosion is set up which is difficult to stop.

Figure 7.10   Causes of soil erosion, most typically in Third World countries such as Nepal and Ethiopia (Source: Hartzell, 1986)

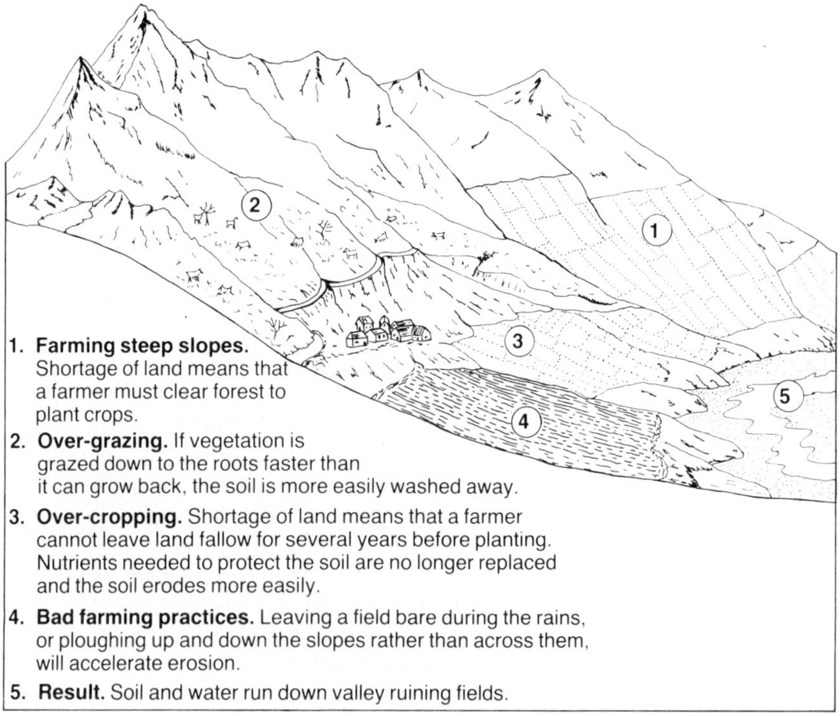

1. **Farming steep slopes.** Shortage of land means that a farmer must clear forest to plant crops.

2. **Over-grazing.** If vegetation is grazed down to the roots faster than it can grow back, the soil is more easily washed away.

3. **Over-cropping.** Shortage of land means that a farmer cannot leave land fallow for several years before planting. Nutrients needed to protect the soil are no longer replaced and the soil erodes more easily.

4. **Bad farming practices.** Leaving a field bare during the rains, or ploughing up and down the slopes rather than across them, will accelerate erosion.

5. **Result.** Soil and water run down valley ruining fields.

3. The effects of soil erosion

Soil erosion results in the thinning of topsoil and the loss of plant remains and nutrients from the soil. Crop yields are reduced and the productivity of grassland is lowered. This has the effect of increasing farming costs while lowering incomes. In the poorest nations, farmers find it harder to feed their families. Faced with famine they may try to exploit the land even further by cutting down trees and removing crop-stubble and roots from the land to feed cattle. These plant remains should of course be left in the soil to improve soil structure and to retain water. Another vicious circle of soil erosion thus occurs: farmers, faced with soil erosion, have no choice but to speed it up even more. Other farmers faced with starvation may migrate, often across international borders, to become 'ecological' or environmental refugees.

Another reason why erosion is difficult to control is that many of its most costly and damaging effects occur great distances from the eroding areas. Eroded soil which is washed into rivers increases the amounts of sediment and pollution: flooding, the dredging of ports and harbours, the damage to fish-breeding areas, the silting-up of hydro-power dams, are among the hidden costs of this. In the USA, for instance, the annual cost of so-called 'off-site' erosion is estimated at $6 billion.

4. How to stop soil erosion

A variety of remedies and solutions are available to control soil erosion (see Figure 7.11).

(a) Technical solutions

These fall into three groups.

(i) *Mechanical methods*. One of these is the building of terraces across slopes to hold the soil on the land (see Figure 7.11 and Plate 7.3). Eventually the terrace becomes level as the soil is caught when it washes down. In Britain the rates of erosion are not severe enough for terracing to be needed. In the tropics, however, erosion can be considerable: almost all of the rain falls in fierce storms during the rainy season and causes major soil loss. The disadvantage of this method is that it can take up to 10% of land out of production. Other mechanical methods of protection are the construction of bund-like embankments to retain soil wash, contour ploughing, where the farmer ploughs across slopes rather than along them (see Plate 7.4), and the building of shelter-belts, such as a line of trees, which check wind-speed and help to protect the land from wind-erosion.

(ii) *Crop methods*. This range of techniques focuses on the system of cropping. Care may be taken, for instance, to maintain a crop cover for as long as possible to protect the soil from erosion. Rates of erosion are very low when the soil is covered by grass, so the simplest solution to the problem is to re-introduce crop rotations using grass leys. Also, rather than leaving land bare between the crop-rows, other crops can be grown in the open

Figure 7.11   Methods used to prevent soil erosion, particularly in Third World countries (Source: Hartzell, 1986)

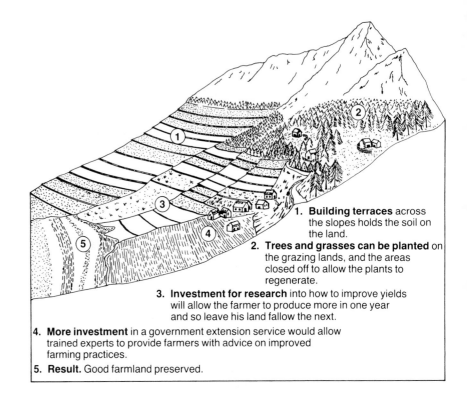

1. **Building terraces** across the slopes holds the soil on the land.
2. **Trees and grasses can be planted** on the grazing lands, and the areas closed off to allow the plants to regenerate.
3. **Investment for research** into how to improve yields will allow the farmer to produce more in one year and so leave his land fallow the next.
4. **More investment** in a government extension service would allow trained experts to provide farmers with advice on improved farming practices.
5. **Result.** Good farmland preserved.

Plate 7.3   Terraced cultivation helps to check the devastating effect of soil erosion in tropical mountain countries, such as here in Nepal. (Photograph: L. Burgess)

Plate 7.4 Fields of wheat stubble, Parana State, Brazil. In this area of heavy rainfall, contour ploughing is practised, and bund-like embankments are built to protect the soil from erosion (Photograph: Tony Morrison).

rows (i.e. inter-cropping), e.g. in the tropics rows of coffee bushes may be inter-planted with maize.

(iii) *Soil methods*. These methods of protection concentrate on the soil system itself. The aim is to reduce erosion by building up and maintaining the fertility and structure of the soil. The most practical way is to add organic matter to the soil. This, as suggested on page 20, will increase the ability of the soil to hold water and will promote a stable soil structure. Two ways of doing this are: (1) growing a grass crop to increase the humus content of the soil; (2) adding crop-stubble and straw as well as farmyard manure. Another soil method is to choose the right kind of tillage which will not pulverise (grind down) and deform soil structures (see page 24). Heavy clay and waterlogged soils are easily damaged during ploughing. These fragile soils benefit from the installation of pipe drains and from the break-up of a compacted soil by *subsoiling*. e.g. deep ploughing that mixes the surface and subsoil horizons.

(b) Political solutions

Although the technical solutions to soil erosion are well known, the political will to implement them is lacking. As a result, little money has been invested in soil conservation by governments and aid agencies, such as the World Bank. In the Third World, more investment for research into how to improve the yield of local crops would allow farmers to produce more in one year and so leave the land fallow to recover the next year. More investment in government extension services would allow trained experts to advise farmers on improved farming techniques. More investment in projects already set up to encourage poor farmers in soil conservation would provide more incentives to help these villagers to keep up the increased yields: new roads, transport, guaranteed market prices and security in land-holdings are just as important as the technical solutions.

In the rich countries, however, the political solutions to soil erosion are rather different: the problem is that too much incentive is given to farmers, in the form of guaranteed prices for crops, loans and tax-free concessions, to increase their outputs from the land. Farmers need to be encouraged to produce less not more, while keeping up their livelihood. Only in this way will the dangers of modern intensive farming (see pages 22–4) be reversed.

5. Case studies in soil erosion

(a) Nepal

(i) *Scale of destruction*. Nepal, situated on the southern flanks of the Himalayas, is one of the poorest nations on earth, with an average yearly income of only about £120 (*cf*. UK £6208) per person. Most of its people have to grow enough food each year in order to survive. It is particularly tragic, therefore, that soil erosion has reached crisis-level. The reasons are not hard to find. In recent years, partly as a result of a high rise in population, much of this mountainous country has been rapidly cleared of forests. When the rains come in the summer monsoon, they fall in great torrents. Their energy easily breaks loose the unprotected soil surface, dislodges the soil particles and carries them off down slopes and streams.

As more forests are cleared by farmers, the topsoil, which is the wealth of the land, is eroded and so harvests fail. In the Nepalese hills, rice yields have dropped by 20% in five years, maize yields by 33%. With declining crop yields, farmers become poorer and they must exhaust the land still further to survive. This vicious circle, in which poverty and soil erosion chase each other, usually ends in starvation.

Enough soil is lost every year from the fields of Nepal to form whole new islands downstream in Bangladesh. Where the rivers leave the foot-hills of the Himalayas, they slow down and deposit their sediment over the great flood-plains of the River Ganges, known as the *terrai*. In the process, the rivers change course violently, resulting in flooding and deaths by the thousand.

In Nepal, in order to try to cope with the threat of soil erosion on steep slopes, terraces are cut in the mountain sides (see Plate 7.3). But the soil erosion is now so great that whole terraces and hillsides can be washed away. Each time the monsoon rains come, scores of people are killed in landslides and hundreds of hard-won terraces are destroyed.

(ii) *Farming system and distribution*. Worst affected by erosion are the middle hills of Nepal situated between the lower foothills and the Himalayan Mountains. The forested slopes too steep to cultivate (see Figure 7.10) are especially vulnerable. Most people are supported by tiny farms perched on the shoulders of the mountains. In this region populations have doubled within a generation, so that eight million people now live in the middle hills. As the protecting cloak of forest has been cut more and more, the landscape has crumbled around them. Famine now strikes deep into the land.

Before the recent rise in population, farmers and their families did survive in this region, although it is a fragile environment. The tiny farms of the middle hills contain on average five persons and are about one-third of a hectare in extent. A typical system is this: farmers grow maize and millet; the stalks of these crops are fed to buffalo which provide milk and meat and power in ploughing; the buffalo also convert the stalks to manure, which is spread on the land to return valuable nutrients that the crops have removed (see Figure 4.9 on page 103). Without cattle and buffalo the tiny farms would lose their fertility.

Apart from the terraced fields, the surrounding grazing and forested land is important to the stability of the system. Trees are cut and used as fuel for cooking. Cattle are fed grass and tree leaves, and dung from their pens is returned to the terraced fields. For the system to be in balance, there needs to be one cow or buffalo for each person and they need to graze four times as much forest and grassland as each family cultivates. Now with increasing population there are too many mouths to feed. There are too many cattle for the grazing and forest land: the grass is over-grazed and the ground is laid bare to the rains; the forest is destroyed by cutting and young tree seedlings are eaten by goats.

(iii) *Community forestry in Nepal*. The cycle of poverty and land degradation in Nepal is being slowly checked by government programmes of tree planting. Planting schemes were begun in the 1950s but the early programmes were not too successful because the government did not involve the local people. The government fenced off plantations (and thus land) to stop animals from eating the young tree seedlings. However, they lacked the resources to patrol the plantations, which were soon invaded by the locals and were destroyed.

Since the government has begun involving local community councils in the management of the forests, replanting has been far more successful. Expensive fences are no longer needed as the people protect the young seedlings themselves. Rather than let their animals roam in the plantations, where the seedlings would be eaten, the local people are allowed to harvest grass and other vegetable material from the plantations. Animal forage is removed and taken to the village where it is fed to the animals which are kept in pens. Their dung is used to fertilise the land. Such local community forestry is still in its early stages in Nepal and progress can be slow. Although there are certain districts where planting is taking place faster than trees are being felled, in the country as a whole the forests are still being cut down three times as fast as they are being planted.

(b) Britain

It is widely held that soil erosion in Britain is not a problem because of the low intensity of the rainfall, the high proportion of land under grass, and the seemingly widespread distribution of fairly deep and fertile soils. However, soil erosion is a good deal more widespread than we may have thought, and it is getting worse. In some places it presents a serious threat to farming.

The intensification of agriculture and the loss of habitat, outlined above in Section A, have been accompanied by an increased rate of soil erosion. Soil erosion, especially by wind, has been accelerated by the misuse of light sandy soils, the removal of wind-breaks, and the ploughing up of unsuitable soils (notably moorland and drained marshes) for growing cereals. Water-erosion is also widespread where land has been laid bare for arable cultivation and where steep slopes have been ploughed up.

Damage is occurring particularly on the lighter sandy and chalk soils of Britain, where the topsoil is generally thin (only 2–3 cm in some sandy soils). Recent work by the Soil Survey of England and Wales has shown that, of seven chalkland sites in Berkshire, Hampshire and Sussex, erosion has lowered the ground surface by 6.25 cm in the least affected case and by as much as 21.25 cm in the worst. Where the chalk-down grassland has been ploughed up for agriculture, there has been a serious loss of topsoil.

The effects of erosion can be seen in spring when lighter-coloured patches appear on hillsides, where ploughing has turned the subsoil on to the surface. On the Sussex Downs, between Brighton and Eastbourne, and along the South Downs Way, the chalk is plain to see in some fields. In places most of the topsoil has already gone. In areas where light sandy soils are eroding, deposits of fine material sometimes up to one metre thick can be found down the slopes from fields or in valley bottoms. As soils become thinner and as more and more subsoil is exposed on the slopes above, these deposits become coarser, and the lower land is covered by infertile sandy material.

It has been estimated that about 1 million hectares or about 18% of the arable area of England and Wales is at risk from high rates of soil erosion. This figure refers only to land suitable for agriculture. It does not include the misuse of land where, for example, land of lower class is used for arable cultivation. Such land includes, of course, many of the light sandy, peaty and chalk soils already noted as under threat. Many of these vulnerable areas might be turned into wastelands if present farming practices continue. Within fifty years many of them will not be able to be farmed.

ASSIGNMENTS
1. (a) *Define soil erosion, and describe the main agents responsible for it.*
    (b) *Refer to Figure 7.10 and Plate 7.2. What is accelerated soil erosion and how is it begun?*
    (c) *Describe the 'on-site' and 'off-site' consequences of soil erosion.*
    (d) *Explain why soil erosion is difficult to control. In your answers, use: (i) the idea of vicious circles; (ii) off-site erosion effects; (iii) the role of government.*
2. *Refer to Figure 7.11 and Plate 7.3.*

(a) *How can soil erosion be controlled by improving: (i) the physical shape of the environment; (ii) the cropping system; (iii) the soil condition?*

(b) *Why might such technical solutions fail if they have no political backing?*

3. (a) *Examine the roles of climate, terrain and removal of vegetation in soil erosion in Nepal.*

(b) *Describe the local and regional effects of soil erosion in Nepal.*

(c) *Show how the farming system in Nepal: (i) formerly maintained a stable soil system; (ii) now, as a result of population increase, causes severe soil erosion.*

Key Ideas

A. *Britain's Lost Landscapes*

1. Modern intensive agriculture has been responsible for destroying many of Britain's semi-natural habitats together with their valuable, associated wildlife.
2. The habitats most affected in the last 40 years include unimproved meadow, chalk downland, heathland, ancient woodland and lowland bog and marsh.
3. Associated species may become extinct or be on the verge of extinction (i.e. endangered) by the destruction of these habitats.
4. Loss of habitat has caused the extinction of the Large Blue Butterfly and has endangered the existence of plants such as the Monkey Orchid and the Marsh clubmoss.
5. Over the last 40 years, a number of areas have been chosen in which to conserve habitats and their wildlife.
6. They include National Parks, Areas of Outstanding Natural Beauty, National Nature Reserves and Sites of Special Scientific Interest.
7. The designation of such conservation areas is, however, no guarantee of their protection, and much damage can still occur within them.

B. *Acid Rain Over Europe*

1. Acid rain is a general term used to denote the total deposition at the earth's surface of a wide range of acidic pollutants from the atmosphere.
2. Acidic pollutants, e.g. sulphur dioxide and nitrogen oxides, can be deposited by rain (hence the term), by snow and mist, by wind, and by gravity.
3. Acid rain can be transported by wind across national boundaries.
4. For instance, Britain is a major producer and exporter of acid rain, while Sweden is a minor producer but a major importer.
5. Acid rain has been blamed for making acid many rivers, lakes and soils and for the consequent harmful effects on wildlife.
6. Many forests in Europe are dying because of acid-pollution fallout.

7. Many countries in Europe, except Britain, are attempting to reduce the effects of acid rain over the next several years.

C. *The World's Vanishing Soil*

1. Soil erosion occurs when soil is removed from an area by wind, water or gravity.
2. Soil erosion is a natural process but humans greatly accelerate it by misusing and over-exploiting the land.
3. Accelerated soil erosion occurs when the structural strength of a soil is reduced by a loss of soil organic matter and/or when the protecting vegetation cover is removed.
4. Soil erosion can cause 'on-site' damage, such as crop failure and famine, and 'off-site' damage, including build-ups of river sediment, pollution and flooding.
5. A range of technical solutions is available to control soil erosion.
6. These include methods which focus on improving: (i) the physical form of the environment; (ii) the cropping system; (iii) the soil condition itself.
7. Soil conservation programmes are little developed, however, because they often fail to receive government support.
8. Heavy rainfall acting on steep slopes stripped of trees has resulted in severe soil erosion in Nepal.
9. Community forestry programmes in Nepal may help to check the rate of soil erosion.
10. Soil erosion in Britain is occurring faster than is generally believed.
11. It is a cause of some concern, particularly with regard to light soil used for intensive agriculture.

Additional Activities

1. (a) Examine the following groups of organisms: (i) apes and chimpanzees; (ii) kangaroos; (iii) primroses and butterflies; (iv) whales and herrings; (v) colonial sea birds; (vi) wild varieties of wheat.
   (b) Give reasons why you think they should be conserved, using one or more of the following reasons: for aesthetic pleasure; because they are of economic value; because they are of value in the studies of humans and of behaviour; as living 'fossils' of use in the study of evolution; as a moral obligation; because they are of use in the study of population ecology.
2. Pay a visit to your nearest local or national nature reserve and then do the following:
   (a) Outline the geology, relief and climate of the reserve.
   (b) Make a list of the main habitats and species.
   (c) Describe the main pressure faced by the reserve and its species.
   (d) Outline the main methods used in the conservation of the different habitats and their associated species.

Table 7.1 Average annual soil losses in the Silsoe area, Bedfordshire

Plant cover/soil texture	Slope	Rate of erosion (tonnes/ha)
Bare sandy loam soil		10–45
Cereals on sandy loam soil		0.6–24
Cereals on chalky soil	7°–11°	0.6–21
Cereals on clay soil		0.3–0.7
Grass on sandy loam soil		0.1–3.0
Woodland on sandy loam soil	20°	0.01
'Acceptable' rate		0.1–1.0

Source: Morgan, 1986

3. Refer to Table 7.1.
   (a) Describe the pattern of soil erosion shown.
   (b) Examine the role of (i) soil texture and (ii) the value of a plant cover, in influencing the rate of soil erosion.
   (c) What does this table suggest about the future condition of the light arable soils of England and Wales?
4. Refer to Figure 7.12.
   (a) Describe the patterns of acid precipitation shown in Figure 7.12.
   (b) Describe the patterns in lake acidity shown in Figure 7.9.
   (c) Describe the trends in soil acidity shown in Figure 7.8.
   (d) Examine the relationships between the patterns shown in Figure 7.12 and those in Figures 7.8 and 7.9.
5. Examine Figures 7.13, 7.12 and 2.21. Using a good atlas, draw maps of Great Britain showing the distribution of: (i) solid rock geological formations; (ii) amounts of rainfall. Now do the following:
   (a) Describe the distribution of the main rock formations in Great Britain. Note that hard igneous rocks such as granite are not able to neutralise acid rain while chalk and limestone can.
   (b) Using Figure 2.21, describe the distribution of soil acidity in Great Britain.
   (c) Using Figure 7.12, describe the distribution of acid rain deposition in Great Britain in 1974.
   (d) Summarise the pattern of acid-rain damage shown in Figure 7.13.
   (e) Examine the relationship between the pattern of acid-rain damage and: (i) the 1974 distribution of acid rain; (ii) the distribution of amounts of rainfall (note that the more rain that falls, the greater will be the total deposition of acid rain); (iii) the distribution of acid rocks and soils
   (f) Which factor is the most important in determining the pattern of acid-rain damage?

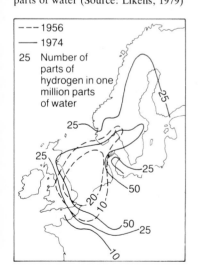

Figure 7.12 Acid precipitation in north-west Europe, 1956 and 1974. The units show the concentration of hydrogen in rain and snow, measured as parts of hydrogen in one million parts of water (Source: Likens, 1979)

--- 1956
— 1974
25 Number of parts of hydrogen in one million parts of water

Figure 7.13   Acidification of water systems in Britain and the effect on wildlife. Evidence points to declining numbers of all these species of flora and fauna in areas vulnerable to acidification. (Source: Lean, 1986)

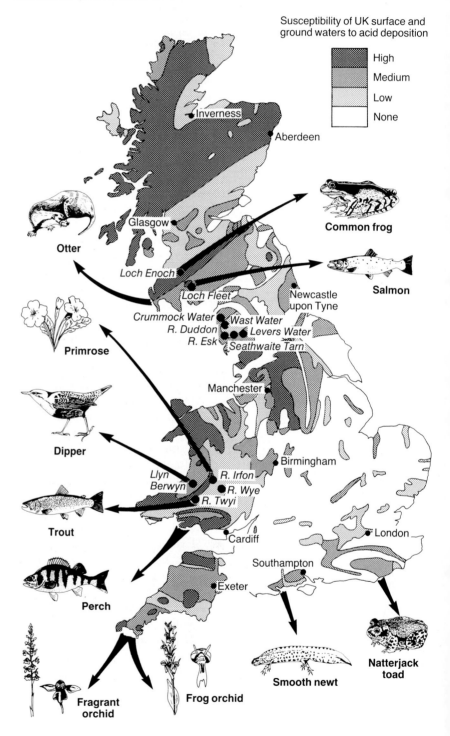

8 Classroom Analysis of Soil Properties

Soils and their properties can be more fully understood by collecting soil samples in the field and taking them back to the classroom or laboratory for analysis. When you come to measure the various soil properties outlined in this chapter, your ability to do so and the validity of your results will depend on the equipment available to you. You will find that most soil analysis is better carried out in the chemistry laboratory than in the geography classroom. For successful results, you must consistently follow certain analytical techniques and procedures which are outlined below in Sections A and B. Finally, remember that soil analysis is time-consuming. While it may take only an hour or so to collect local soil samples, their physical and chemical analyses may take 12–24 hours!

A. Sampling Techniques

When soils are sampled in the field, a sufficient amount of soil must be collected. For instance, an individual soil sample may be tested for more than one soil property and so the original soil sample has to be divided into several smaller sub-samples. Also, it may be useful when measuring a soil property to make two, three or more separate measurements of the soil sample so that a more reliable, 'average' calculation can be made. For these reasons, 50–100 g of soil is normally collected at each sampling-point. The collected samples can be placed in polythene bags, sealed and then labelled according to their general location in the field (e.g. valley site, mountain slope, school grounds, local garden). Their position and depth in the soil profile (e.g. A-horizon at 5–10 cm, B-horizon at 15–20 cm, C-horizon at 50 cm) must also be recorded. Such standardisation of position and depth of sampling is necessary if soil samples from different locations are to be directly compared.

Soil can be collected either as a *disturbed* sample, where a rough amount of soil is simply trowelled into a polythene bag, or as an *undisturbed* sample, where a known volume of soil is collected by using a core sampler. A shallow piece of metal pipe, 4–5 cm in diameter and 2.5–5 cm in depth can be used for this purpose. This open pipe can be pressed or hammered into the soil to enclose a column of undisturbed soil. When extracted, the volume of the soil sample is that of the internal volume of pipe, and is given by the formula $\pi r^2 h$, where $\pi = 3.146$; $r = $ radius of pipe; $h = $ height of pipe. The soil from the core sampler may then be 'disturbed' and placed in a polythene bag.

B. Physical and Chemical Analyses

1. Disturbed sample

Disturbed samples are used in a *fresh* state (i.e. as collected in the field) for measurements of pH and soil moisture. Most other measurements of soil properties are based on the *air-dried* condition of disturbed samples.

(a)   Acidity or pH

(i) *Apparatus.* The apparatus needed includes a 50 cm^3 glass beaker, a glass stirring-rod, a wash-bottle containing distilled water, and a pH-meter.

(ii) *Procedure.* Mix approximately 5 g of fresh disturbed soil in about 10 cm^3 of distilled water in a 50 cm^3 beaker. Stir with a glass rod (or shake well) and allow to settle for a few minutes. Measure the pH by introducing the glass probe of the meter into the soil solution. Take care not to push the delicate (and expensive) electrode, at the end of the probe, into the soil at the bottom of the beaker. (Alternatively, soil pH may be measured by using a BDH kit. Directions are usually supplied with this type of apparatus.)

(b)   Moisture content

The moisture content of a soil is defined as the ratio of weight of water to the weight of solid particles in a given volume of soil. The soil sample is weighed, dried at a temperature of 105–110 °C and weighed again. The loss in weight represents the weight of water in the sample.

(i) *Apparatus.* This includes a glass weighing-bottle or suitable container, an accurate weighing balance or scale capable of weighing to 0.1 g or better, and a small oven.

(ii) *Procedure.* Weigh a clean, dry weighing-bottle or container (W_1). Place a sample of about 25 g of loose, fresh soil in the container. Weigh the container and moist soil (W_2). Place container in an oven and dry the soil at a temperature of 105–110 °C for at least 24 hours. After removal and cooling, weigh container and dry soil (W_3). Calculate the moisture content (W), as a percentage of the dry-soil weight, using the formula

$$W = \frac{W_2 - W_3}{W_3 - W_1} \times 100$$

(c)   The air-dry fine earth fraction

This fraction of the soil (i.e. size-class of particles) is used for the majority of physical and chemical analyses that are carried out.

(i) *Apparatus.* Mortar and pestle, 2 mm-gauge sieve with lid and receiver pan, soil tray.

(ii) *Procedure*. Place an approximate 25 g soil sample on a tray in the laboratory. Allow to dry to constant weight at about 25 °C. Empty the air-dried soil sample into a mortar and, using a pestle, *gently* disperse the soil aggregates. Pass the soil through a sieve with holes 2 mm in diameter. The sieve should be shaken for about two minutes, and the soil which passes through the sieve collected in a receiver pan. The soil in the receiver pan is the *air-dry fine earth fraction*. This may be emptied into a soil tray for further analyses. The particles retained in the sieve, including large roots, large pieces of dead organic matter and large pieces of inorganic material (i.e. stones and gravel) are discarded.

(d) Organic matter content

(i) *Apparatus*. Bunsen burner, small procelain crucible, accurate weighing scales.

(ii) *Procedure*. Weigh out enough air-dry fine earth fraction (W_1) into a small procelain crucible so that it is about two-thirds filled with soil. Weigh the crucible and the air-dry fine earth sample (W_2). Place the crucible on a metal stand over a bunsen flame for 15 minutes. This burns off the organic matter in the soil. Weigh the crucible and the resulting mineral soil (W_3) when cool. Calculate the organic matter, as a percentage of the air-dry fine earth fraction, by using the formula

$$\text{percentage organic matter} = \frac{W_2 - W_3}{W_1} \times 100$$

The mineral percentage content of the soil is easily calculated by using the formula $100 - $ (percentage organic matter content).

(e) Carbonate content

Drop a small amount of dilute hydrochloric acid (10% HCl) on to a 10 g air-dry fine earth soil sample. Place the sample near the ear. Listen for, and observe, the degree of effervescence. The concentration of carbonate in the soil can be estimated by using Table 8.1.

Table 8.1 Visible and audible testing of carbonate in soil, using dilute hydrochloric acid (10% HCl)

Description of soil	Calcium carbonate (percentage content)	Test result (using 10% HCl)
Non-calcareous	0–< 0.1	No reaction
Very slightly calcareous	0.1–< 0.5	Faint effervescence just audible
Slightly calcareous	0.5–< 1.0	Moderate effervescence audible
Calcareous	1–10	Small bubbles easily audible
Very calcareous	> 10	Large bubbles easily audible

2. Undisturbed sample

This type of sample (see Section A) is necessary for working out the relative volume of air spaces to solid particles in the soil. In other words, it is used to assess the density of solid particles in the soil.

Soil bulk density

Soil density or bulk density of air-dry soil can be calculated as follows. Allow the soil sample from the core sampler to dry to constant weight at 25 °C in the laboratory. Using the volume of this sample as the volume of the core sampler, the soil bulk density may be calculated by using the formula

$$\text{Bulk density} = \frac{\text{Air-dry weight of soil}}{\text{Volume of soil (cm}^3)}$$

3. Soil Texture

(See Chapter 1, Section B.2.)

(a) Mechanical methods

The technique used will depend on the nature of the soil and the equipment available. If you are dealing with sandy and stony soils, and have access to a system of fine-mesh sieves, then what is termed the dry-sieving method may be appropriate (see next section). Wet-sieving techniques are suitable for determining the texture of silty and clay soils, the individual particles of which are too fine to be mechanically retained, in their dry state, by the sieves. Wet-sieving is based on the different rates of sedimentation (settling out) of fine soil particles in a column of water. It is a fairly time-consuming and complicated technique, however. A useful outline of the technique and the apparatus required is given in *Soils* by D Briggs (1977).

(b) Dry-sieving

(i) *Apparatus*. Oven, pestle and mortar, accurate weighing scales, nest of fine-mesh sieves and receiving pan (see Plate 8.1), soil tray. The following mesh-diameter sieves, if available, will trap a range of different-sized soil particles: 2 mm (stones and gravel), 1.00 mm (very coarse sand), 0.5 mm (coarse sand), 0.25 mm (medium sand), 0.125 mm (fine sand), 0.0625 mm (very fine sand), 0.031 mm (coarse silt), receiving pan (medium and fine silt plus clay).

(ii) *Procedure*. Dry a reasonably large soil sample (about 50 g) in an oven at 105–110 °C to constant weight. Break up the soil sample, if necessary using a pestle and mortar, and *gently* grind the soil aggregates. Do not crush them. Set the sieves which are available in sequence, putting the coarsest mesh at the top of the set. Place a closed pan at the

Plate 8.1 A set of soil sieves and their different-sized mineral fragments or classes. The mesh diameters of the sieves are, from finest to coarsest, 0.063 mm, 1.0 mm, 2.00 mm and 6.7 mm. (Photograph: G. O'Hare)

bottom of the set. Pour the sample soil (in stages if necessary) into the top sieve and place the cover on. Lock the sieve set on a sieve shaker (if available) and switch on the shaker for 10 minutes. Otherwise, manually shake the set of sieves for 15 minutes. Select clean, dry plastic trays or other suitable receptacles for the sieved remains. Weigh the trays accurately to at least 0.1 g. After shaking, carefully empty the contents of each sieve pan into a pre-weighed tray. Tap each sieve rim on to a piece of paper to remove any obstinate grains and place them in the trays. Do *not* gouge at the sieve-mesh nor use a hard brush on the finer-mesh sieves. Weigh each tray plus each of the sieved remains accurately, to at least 0.1 g. Subtract the previously weighed tray weight, to give the absolute weight of sediment for each sized grade. The weights of the various size-classes can then be worked out as percentages.

(c) *Determining texture by feel*

This procedure consists merely of rubbing the soil between the thumb and forefinger. It can be carried out in the classroom or in the field (see Chapter 9). As shown in Figure 8.1, *sands* and *sandy loams* are distinctly gritty to the touch and become finer as a *loam* is reached. Most of the particles can be seen without the aid of a magnifying glass. When moist or moistened, they cannot be rolled into thin rods or ribbons between the palms of the hands. Soft clods form when these soils are dry. *Clay*

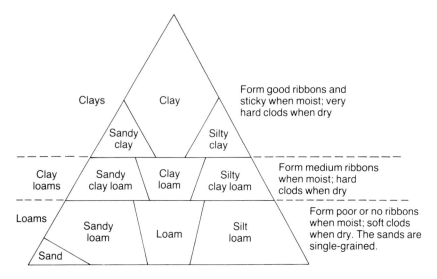

Figure 8.1 Modified textural triangle for determining soil texture by the 'feel' method

loams form medium rods when moist and hard clods when dry. *Clay soils* form good rods and are sticky when moist. Very hard clods occur when clay soils dry out.

The following definitions of the major textural groups may be useful for more detailed soil analysis.

(i) *Sand*. Sand is loose and single-grained. The individual grains can readily be seen or felt. Dry sand, if squeezed in the hand, will fall apart when the pressure is released. If squeezed when moist, it will form a cast but this will crumble when touched.

(ii) *Sandy loam*. A sandy loam is a soil containing much sand but which has enough silt and clay to make it stick together a little. The individual sand grains can be readily seen and felt. If squeezed when dry, it will form a cast which will readily fall apart; if squeezed when moist, a cast can be formed that will bear careful handling without breaking.

(iii) *Loam*. A loam is a soil which has a fairly even mixture of the different grades of sand, silt and clay. It feels somewhat gritty, yet fairly smooth and slightly plastic. If squeezed when dry, it will form a cast that will bear careful handling; if squeezed when moist, the cast formed can be handled quite freely without breaking.

(iv) *Silt loam*. A silt loam is a soil with a moderate amount of fine grades of sand and only a small amount of clay. Over half of the particles are of the size called silt. When dry it may feel soft, smooth and floury. Either dry or moist, it will form casts that can be freely handled without breaking; if squeezed between the thumb and finger when moist, it will not form a ribbon but will give a broken appearance.

(v) *Clay loam*. This is a fine-textured soil, with features midway between the characteristics of a loam and a clay. If moist, it is possible to form a ribbon with it when squeezed between the thumb and finger.

(vi) *Silty clay loam*. A silty clay loam is a fine-textured soil which breaks into clods and lumps. When these are dry, they are hard to break with the squeeze of the hand. When the moist soil is pinched between thumb and finger, it will form a cast that will bear much handling. When kneaded in the hand, it does not crumble readily but tends to work into a heavy, plastic, compact mass.

(vii) *Clay*. A clay is a fine-textured soil that usually forms very hard lumps or clods when dry, is quite plastic and is usually sticky when wet. When the moist soil is pinched out between the thumb and fingers, it will form a long, smooth, flexible ribbon.

C. Projects

1. Comparative analysis of soil texture

(a) Using a 1:63 360 or 1:25 000 Soil Survey map of your local area, identify six soil types of widely differing texture.
(b) Collect one moist soil sample of each type. Bring the soil samples, which need to be about 500 g each, to the classroom in polythene bags.
(c) Sub-divide the samples and ask students to estimate the percentage of sand, silt and clay of each soil sample. This can be done by identifying soil texture, using the descriptions given on page 180. The estimated classes of soil texture can then be related to the textural triangle in Figure 8.1 to provide approximate percentage figures.
(d) Tabulate the sand, silt and clay estimations of each student for the various soil samples.
(e) Discuss the spread of results, noting which soils seem to be more easily determined for texture than others.
(f) If you have the equipment available, you can use dry- and wet-sieving methods to determine the soil texture of the samples more accurately. Such mechanically determined results can then be compared with the students' estimations.

2. The effects of trampling on soils

(a) *Aims*

To investigate the effects of different degrees of trampling on: (1) soil moisture content; (2) soil pH; (3) content of soil organic matter; (4) soil bulk density.

(b) *Procedure*

Select a suitable 'natural' path from which much vegetation has been removed, e.g. by the side of the school field, or a road verge.

Make an accurate sketch of the path, carefully recording the dimensions of bare soil and vegetation. Choose two locations. Ideally these should illustrate the different degrees of trampling, such as one site at the narrow end of the path (heavy pressure) and one site at the wide end (moderate pressure).

At each of these two locations, collect the data shown below from: the middle of the path, the edge of the path where more vegetation is present, and just off the path where there is much vegetation and little sign of trampling.

(i) At each of the six sites, use a trowel to collect a 50 g sample of soil. The soil samples should be taken from a standard depth, i.e. at about 5 cm from the surface. Analyse the soil samples for moisture content, pH, and content of organic matter, using the techniques described in the previous section.

(ii) At each of the six sites, collect a surface sample of soil by using a core sampler. You may need to remove surface vegetation and obvious surface organic matter before doing this at the path-edge and off-path sites. Determine the degree of soil compaction by calculating the soil bulk density at each site. Use the method explained on page 178.

Soil compaction may be estimated if you do not have a core sampler: hold out a metal stake at arm's length above the soil sites and let it fall through your fingers. Measure the depth of entry into the soil as an indicator of soil compaction. This may seem unscientific, but it really does work if you take care to hold the stake at a constant level for each drop.

(iii) Tabulate your results to enable you to draw comparisons between the three recording sites at each location and between the two locations.

Discuss the relationship between the measured soil variables and: (i) the apparent degree of trampling; (ii) the amount and degree of vegetation cover.

Field Observations of Soil Profiles

A. Making a Start

1. Ground rules

If you decide to undertake a project based on the observation and recording of soils in the field, the following ground rules may be of some use.

(a) Objectives

Develop some clear ideas about what sorts of relationship you wish to investigate *before* you venture into the field. This will not only save time later on but also allow you to choose your survey area and soil-recording sites.

(b) Background information

It is useful to have information on what types of soil are found in an area, so that you can survey soils that are typical of that location. Do not spend time on minor, exceptional or unrepresentative soils. Make use of relevant topographic, geological and soil maps. Soil maps at scales of 1:25 000, 1:63 360 and 1:100 000 are published by the Soil Survey and cover about one-quarter of the land of England and Wales. Some of these maps are accompanied by Soil Monographs which describe the geology, climate, land-use and soils of the area.

If your survey area is not covered by the above maps and their monographs, you can refer to the Soil Map of England and Wales (1:250 000). This map, published in 1983, is produced in six sheets (see Figure 9.1). Each sheet is accompanied by a comprehensive Regional Bulletin (1984), describing the soils of the district. There is also a smaller Legend booklet (1983) which briefly outlines the constituent soils. This booklet contains a useful description of the way in which the mapped soils of England and Wales are presently classified into ten major groups, including brown soils, podzols, peat soils, etc (see Avery, 1980).

These maps and bulletins and other Soil Survey publications are available from the Publications Officer, Soil Survey of England and Wales, Rothamsted Experimental Station, Harpenden, Hertfordshire AL5 2JQ. Soil maps and monographs covering Scotland (see Figure 9.1) can be obtained from The Macaulay Institute for Soil Research, Craigiebuckler, Aberdeen AB9 2QJ.

Figure 9.1   Index to the 1:250 000 Soil Survey maps of (a) Scotland, (b) England and Wales

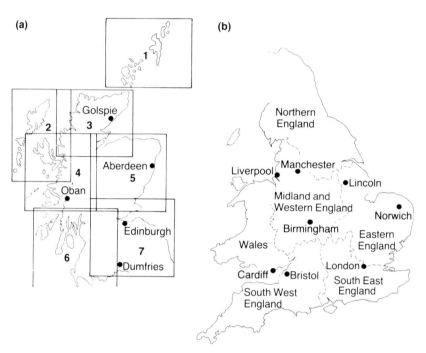

(c)   Demands of time

Remember that the field observation of soils (as with laboratory analysis) is fairly time-consuming. Do not expect to investigate half a dozen soils during one field visit. It may take you a whole day to travel to the survey area, dig a soil pit (see Section 2 below) and then describe one, perhaps two, soils!

(d)   Equipment

Useful equipment for soil description includes a sturdy steel spade, a trowel, a metre rule, and a camera if possible. Good colour photographs of soil profiles and their immediate environment save long soil and site descriptions. They also provide an image of the total soil profile which cannot be explained by words alone.

(e)   Combating weather

Be prepared for inclement weather, especially in exposed upland sites, where atmospheric conditions can deteriorate rapidly. Take a pair of strong boots (for walking *and* digging) and both warm and waterproof clothing. It can be warm digging a soil pit, but cold when recording soil profile data!

Bogged down in
soil studies?

(f) Field etiquette

Never dig a soil pit without first obtaining permission from the landowner or manager, whether a farmer, the Forestry Commission, National Park Authority, etc (see Figure 9.2). It is also sound practice to write a letter of thanks after the field survey has been completed. Other students may wish to work in the area at a later date, so create a good relationship between your school or college and the proprietor of the land.

2. The soil pit or cutting

(a) Shortcuts

In some areas digging a soil pit can be avoided by using soil cuttings. These are ready-made soil sections produced by natural erosion, land-sliding, road cuttings, etc. If you wish to carry out a rapid general survey of soil properties over an area, soil augers can be used. These are T-shaped extendable metal rods with a screw-head for boring into the ground. They are used to extract a column (sample) of soil which can be rapidly tested for such properties as depth of topsoil, depth of bedrock, texture of parent material, horizon colour, etc. Their limitation is that the soil samples which are collected in the screw-head are usually fairly disturbed.

(b) Digging a soil pit

Describing a soil profile in the field means having to dig a soil pit or cutting. As shown in Figure 9.3, this involves excavating a plot of land about 75 × 75 cm in surface area and, depending on local soil conditions, of varying depth. The pit may be stepped as shown. Make sure that the

Figure 9.3 Size and orientation of the soil pit. These dimensions are ideal but smaller, less laborious pits can often give successful results.

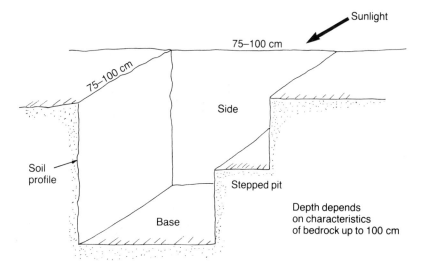

185

soil face you have cut is facing towards the sun. Good light is necessary for detailed observation and, of course, for taking profile photographs.

It is essential that you 'clean up' the soil profile by scraping the face of the surface with a trowel. This makes clear the horizon sequences and often reveals patterns which were not apparent before cleaning.

Finally, you must not leave gaping holes in the countryside, as they not only are a danger to animals and humans but also look unsightly. So please, refill the excavation plot, putting back carefully any grass turfs that you may have disturbed.

B. Making Records

When describing soil profiles in the field, it is important to make as clear and as accurate a record of your observations as possible. You can do this by measuring soil data in a systematic and standardised way, as the notes which follow show. A list of easily observed soil properties is given, together with how they can be measured.

1. Soil and site description sheets

As well as making written records of your field observations, information can be entered onto a soil site description sheet of the type shown in Table 9.1 or onto a soil profile description sheet (see Table 9.2). These data sheets should contain the 'essentials' of your observations.

Table 9.1   Soil site description: student's record

SOIL SITE DESCRIPTION	
Name *Pete Bog*	
Soil reference/number: *PB/4*	Date: *4 April 1987*
Map reference: *SX 639758*	
Profile type: *Peaty iron pan podzol*	
Locality: *Belever tor, Dartmoor*	
Landform type: *Concave valley slope*	
Elevation: *360 m*	Aspect: *South-east*
Exposure: *Moderate to severe*	Slope: *4°*

Drainage: Regional *West Dart drainage basin*
      Surface *Shedding site*
      Profile *Poor above iron pan, well drained below*	

Parent material: *Weathered granite* Vegetation and land-use: *Purple moor grass, sheep grazing* Climate and recent weather: *Cold and wet, recent rain*	

Figure 9.4   Soil profile diagram of a peaty iron pan podzol, Dartmoor

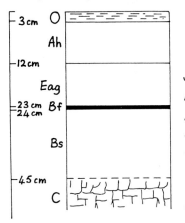

2. Field observations

The following field observations can be entered onto the soil site description sheet.

(a)   *Soil reference/number*

A convenient reference/number system uses the observer's initials plus the number, in sequence, of the observations. For example, if your name is Pete Bog and the soil is the twenty-first you have recorded, then the reference number is PB/21.

(b)   *Map Reference*

Enter the Ordnance Survey full grid reference together with the map sheet number.

(c)   *Profile type*

You can enter the classification once you have completed the soil profile description (see later, *Final classification*).

(d)   *Locality*

Give the name of the general district in which the soil is located, e.g. North Downs, Peak District, Morecambe Bay.

(e)   *Relief characteristics of site*

(i) *Landform type.* It is often possible to recognise that the soil being described is sited on a particular landform, e.g. flood plain, drumlin, scarp slope, or river terrace.

Table 9.2 Soil profile description: student's record

SOIL PROFILE DESCRIPTION

SOIL REF PB/4 Horizons — Name and bottom depth	Boundary	(Munsell) Colour	Texture (particle size class)	Stone content	Structure	Moisture and drainage (mottles)	Organic matter	Roots and fauna
O 3 cm	Abrupt, smooth	Dark brown	Dark brown	—	Very fibrous	Poor	Dead grass	Some roots
Ah 12 cm	Abrupt, smooth	Black	—	—	Fibrous	Poor	Peaty mor-humus	No earth-worms; fine grass roots
Eag 23 cm	Abrupt, irregular	Dark grey stained black	Coarse, sandy and gravelly	Few stones	Structureless; individual grains	Poor drainage; red mottles in grey matrix	Humic staining	Few roots
Bf 24 cm	Abrupt, irregular	Orange-brown	Iron concretion	—	—	Impermeable	—	—
Bs 45 cm	Gradual, smooth	Orange-brown; iron staining	Sandy	Some small stones (1cm)	Poorly structured	Good drainage	Some humic staining	No roots
C ?	—	Light brown-grey	Gravelly	Stones (3-5 cm)	Structureless	Good drainage	—	—

(ii) *Elevation*. The height above sea level.

(iii) *Aspect*. This is the direction in which the site *faces*. Using a compass, take a bearing perpendicular to the slope of the land facing the horizon. If no compass is available, then an estimate will have to be made from an Ordanance Survey map. A slope of less than 2 ° (inclination) may be said to have no aspect.

(iv) *Degree of exposure*. Describe as severe (i.e. top of hill), moderate (hill-slope) or low (valley).

(v) *Steepness of slope*. Measure the angle of slope using a clinometer or an Abney level if available.

(f) Drainage

(i) *Regional*. Name the drainage basin within which the site is located.

(ii) *Surface*. Describe the characteristics of the run off, e.g.
Receiving site: Water collects and slowly infiltrates into the soil.
Normal site: Free water stands for only short periods. Good drainage.
Shedding site: Rapid run off from site. Little infiltration.

(iii) *Profile*. Note the depth of the water table, and the presence of permeable or impermeable horizons.

(g) Parent material

State the rock type (e.g. limestone, sandstone, granite) and name (e.g. Upper Chalk, Millstone Grit) on which the soil lies. If possible, state the nature and degree of weathering (e.g. highly weathered, unaltered bedrock) and note the presence or absence of jointing (i.e. lines of fracture in rock).

In many instances the parent material of the soil may not be the solid rock beneath. For example, above the solid rock there may be a layer of glacial till which forms the parent material. Other superficial deposits which may form parent materials include:
1. Alluvium: river-deposited material
2. Colluvium: deposits at the base of slopes and which have moved down the slopes under gravity
3. Lacustrine deposits: deposits on the floor of an old lake or on its margins
4. Marine sediments and beach deposits
5. Loess: wind-deposited fine silt
6. Sand dunes
7. Glacial drift: e.g. glacial till, fluvioglacial deposits
8. Peat: thick layers of undecomposed plant remains

(h) Vegetation and land-use

Name the vegetation (and/or land-use) within 15 metres of the site. List the main species present with approximate cover value. See Chapter 3, section E.

(i)   Weather

Weather can influence the character of a soil: for example, if there is rain immediately before the observations are made, it can make the soil colours darker. It is important, therefore, to record the weather conditions for the 24 hours before the observations are made. For example, state if there has been rain, a drying wind, frost, or a period of hot sun. Also note the conditions at the time of making the observations.

Information on the categories below may be entered in the soil profile description sheet (see Table 9.2).

(j)   Horizons

Name and record the bottom depth of each horizon. Give the horizon letter according to the following classification:

(i)   *Organic and organo-mineral surface horizons*

L	Undecomposed litter
F	Partially decomposed litter
H	Well-decomposed humus layer, low in mineral matter

Litter layers (bracketing F and H)

A	Mixed, mineral-organic layer at or near the surface
Ah	Uncultivated dark-coloured surface layer
Ap	Ploughed layer of cultivated soils
Ag	An A-horizon with rusty mottling, subject to periodic waterlogging
Oh	Peaty A-horizon
O	Thick peat layer accumulated under wet conditions

(ii)   *Sub-surface layer*

E	Eluvial horizons, depleted of clay and/or sesquioxides
Ea	Bleached (ash-like) horizon (podzolised)
Eb	Brown (paler when dry), friable (apt to crumble), weakly structured horizon depleted of clay
B	Altered horizons which may include some illuviated clay; distinguished from the A- and C-horizons by colour and/or structure
Bw	Weakly developed B-horizon
Bt	Horizon containing illuviated clay
Bh	In podzolised soils, the horizon with maximum deposition of humus
Bs	In podzolised soils, the horizon with maximum deposition of iron and aluminium (sesquioxides); usually red.
C	A horizon that is little altered, except by gleying, and is either like or unlike the material in which the overlying horizons have developed.

Bg,   Mottled (gleyed horizons subject to waterlogging)
Cg,
etc
A/C,   Horizons of transitional or intermediate character
B/C

(k)   Horizon boundary

A two-fold distinction has to be made:

(A)   1.   abrupt   < 2.5 cm wide
        2.   clear   2.5–< 6.0 cm
        3.   gradual   6.0–< 12.0 cm
        4.   diffuse   > 12.0 cm
(B)   1.   smooth   if nearly plane
        2.   wavy   pockets are wider than their depth
        3.   irregular   pockets are deeper than their width
        4.   broken   if parts of the horizon are unconnected with other parts

(l)   Colour

If possible, a *Munsell Soil Colour Chart* should be used to describe the colour of each horizon. If no chart is available then a verbal description will have to be made: e.g. deep red, light yellow, dark grey, white, black. The colour of a soil horizon can be recorded visually by smearing a piece of moist soil on a piece of paper.

(m)   Texture
See Chapter 1, Section B.2, and Chapter 8, Section B.3.

(n)   Concretions and stones

Describe any *concretions* (nodules) present in the soil according to their type, size, shape and density. The main types normally encountered are of carbonates, iron and manganese oxide.

     If *stones* are present in any horizon, state quantity, size and shape.

(o)   Structure

Three items must be recorded for the structure of each horizon: its stability, size and type (shape).

(i) *Stability*. This is usually determined by displacing or gently crushing the aggregates and then noting how durable or hardy are the resulting aggregates. This should refer to the state of the soil at the time of normal moisture content.

0   *Structureless*. No visible structures or peds; single grains or massive (uniform) soil
1   *Weak*. Peds barely observable
2   *Moderate*. Well-formed distinct peds that are moderately durable and evident but not distinct in undisturbed soil

3 *Strong*. Durable peds, quite evident in undisplaced soil, which withstand displacement and become separated when the soil is disturbed.

(ii) *Size and shape*. Refer to Figure 1.9 on page 16.

(*p*) Mottles

The nature of the mottling within the soil gives a good indication of the horizon drainage. You need to record the colour and pattern of the mottles.

(i) *Colour*. Use Munsell notation.

(ii) *Pattern*. Three things need to be defined: the contrast between the colour of the mottles and the rest of the horizon; their abundance (quantity); their size.
Contrast. Faint, distinct or prominent
Abundance. Few (less than 2% of the exposed surface) or many (more than 20% of exposed surface)
Size. Fine: less than 0.5 cm along longest dimension
      Medium: 0.5–2.0 cm
      Coarse: more than 2.0 cm long.

(*q*) Organic matter

L   Undecomposed litter
F   Partially decomposed litter, i.e. fibrous
H   Well-decomposed humus layer, low in mineral matter
A   contrast may be drawn between *mull* and *mor*:
Mull   A humus-rich layer of mixed organic and mineral matter, generally with a boundary grading into the underlying mineral horizon
Mor   Unincorporated organic matter that rests, with little mixing, on the underlying horizon. Poorly decomposed, i.e. fibrous.

(*r*) Roots

Give the type (and name) of the plants that are present immediately above the profile.

(i) *Quantity*
Abundant: more than 200 per 100 cm^2 (10 cm square)
Frequent: 20–200 per 100 cm^2
Few: 4–20 per 100 cm^2
Rare: 1–3 per 100 cm^2

(ii) *Size* (diameter)
Large: more than 1 cm
Medium: 0.25–1.0 cm
Small: 0.1–0.25 cm
Fine: < 0.1 cm

(iii) *Depth*. Note the depth to which the roots penetrate.

(iv) *Shape*. Note whether free-growing, distorted, etc (e.g. roots in clay pans).

(v) *Nature*. Note whether woody, fleshy, fibrous, or rhizomatous (i.e. with horizontal stems in soil).

(vi) *Health*. Note whether dead or alive

(vii) *Age.* Note whether old, young, belonging to past or present vegetational cover.

(s) *Fauna*

Examine the surface for heaps of excavated earth or burrows. Note presence/absence of: (i) animal burrows in soil; (ii) earthworms.

(t) *Reaction*

Test with the BDH kit or with a portable pH meter (see Chapter 8).

(u) *Carbonates*

See Chapter 8, Section B.1.

(v) *Final classification*

When you have completed the soil profile description, you should be able to classify the soil, e.g. podzol, brown soil, brown soil with gleying, etc. (See Chapter 2, Section B.) Enter this on the soil site description sheet.

C. Project Ideas

1. Using your results

Soil profile description remains a rather barren exercise unless the results are related to some environmental factors.

(a) *Soils and vegetation*

A popular project is to measure soil properties against vegetation patterns. For instance, it may be possible to investigate soil types in relation to their formation under deciduous woodland (brown soil) and coniferous woodland (podzol). The influence of other vegetation types can be assessed, e.g. blanket bog and peat soil; heather moorland and podzols; wet acidic grassland and peaty gleyed podzols; bracken and acid brown earth. In order to test the relationship between vegetation and soil, make sure that other environmental variables (e.g. geology, slope) are kept as constant as possible. If you allow these factors to vary as well as the vegetation, eventual correlations will be confusing. Soils should therefore be described as having similar parent materials, slopes, climates, etc.

(b) Soils and slopes

As suggested in Chapter 2, Section C.3, soils vary with respect to slope condition. Select a hill-slope of uniform parent material. Dig a soil pit at the top, middle and lower slope and describe each profile. Relate soil characteristics to drainage condition and surface processes, e.g. soil wash, mass movement.

(c) Soil properties and environmental variables

Rather than digging soil pits and describing whole soil profiles, you may wish to investigate the relationship between a selected number of surface soil properties, or between these and a selected group of environmental factors. This study will involve taking measurements (e.g. of depth of topsoil, amount of organic matter, soil pH, moisture content) at a relatively large number of sites (e.g. about 30). At each site, the slope, altitude, plant species present, land-use, etc can also be measured and recorded.

In order to test the significance of the relationship between:
(i) any two soil variables, e.g. soil organic matter and moisture content,
(ii) a soil variable (e.g. soil organic matter) and an environmental factor (e.g. slope),
(iii) two environmental factors (e.g. different plant species),
you can use statistical methods. This is called association analysis. The chi-square method and Spearman's Rank Correlation Coefficient are useful techniques. These and other statistical methods are outlined in *Statistical Methods and the Geographer* by S. Gregory (1976).

Glossary

Acidity or *pH* The negative $\log_{10}$ of hydrogen ion concentration in solution; expressed as a pH level, e.g. a pH of 4 is acid, a pH of 7 is neutral, a pH of 9 is alkaline.

Acid rocks Rocks rich in quartz and acid in reaction.

Adsorption complex See *colloidal complex*.

Anion Negatively charged ion.

Azonal soil Immature soil lacking well-developed profile.

Bases Metallic cations in the soil, e.g. K^+, Mg^{2+}, Ca^{2+}.

Basic rocks Quartz-free rocks containing the mineral feldspar; basic in reaction.

Biomass The total mass of living organisms present in a community at any one moment; expressed as mass per unit area, measured as dry weight or energy value.

Biome A major global ecosystem. Individual biomes contain *climax communities* of plants and animals and are closely associated with *zonal soils* and climatic regions.

Biota All types of living organisms found in an area; hence *biotic factor* which is the effect of living organisms on environment.

Capillary water The portion of soil water, held by surface tension within the pores, which is largely available for plant-root uptake.

Catena A sequence of soils, usually from similar parent materials and of similar age, whose characteristics differ owing to differences in relief (landform) and drainage.

Cation (*base*) *exchange* The chemical replacement of cations (i.e. exchangeable cations) within the soil. For example, a cation such as K^+ held on the surface of clay or a humus particle can be exchanged for another cation (e.g. H^+) from the surrounding soil moisture. Cations can also be exchanged between soil particles/soil solution and plant roots.

Cation (*base*) *exchange capacity* Ability of clay and humus in the soil to hold and exchange base cations with the soil moisture and plant roots; highest in well-decomposed humus, lowest in certain clays, e.g. kaolin, iron oxide.

Chelate Compound formed when metallic cations (especially aluminium and iron) combine with organic acids in the soil.

Chelation The process by which rocks and soil decompose through the action of organic acids. These acids disintegrate the soil clays, forming chelates with the latter's metallic cations.

Cheluviation A term derived from the combination of chelation and eluviation: the removal of iron and aluminium organic compounds (*chelates*) down through the profile by acid drainage waters (*eluviation*).

Climatic climax vegetation Climax vegetation in equilibrium with its prevailing climate.

Climax vegetation (or *plant community*) A terminal or end-stage plant community in equilibrium more or less with its environment.

Colloidal (or *adsorption*) *complex* Close association of very fine humus and clay particles (colloids).

Conglomerates Very coarse-grained sedimentary rocks containing pebbles.

Ecosystem Biotic community of plants and animals and their associated physical and chemical environment.

Edaphic A term relating to the soil.

Eluviation Process of removal or washing out of soil materials from one horizon to another; usually upper to lower in profile. Compare *illuviation*.

Evapotranspiration The loss of moisture from the earth's surface by means of direct evaporation allied with transpiration from vegetation. *Actual evapotranspiration* is the observed or true loss; *potential evapotranspiration* is the theoretical maximum loss, which assumes an unrestricted supply of water to the surface (e.g. by irrigation).

Field capacity Maximum amount of water retained by soil after all *gravitational water* has been drained.

Flora A term referring to the collection of plant species found in an area.

Flushing Process whereby soils become enriched by transported materials, either dissolved chemical salts or rock particles; hence a *flush* is an area of soil enriched in this way.

Food chain The sequence of energy transfer, in the form of food, from organisms in one *trophic level* to those in another, achieved when organisms eat or decompose others.

Gleying The reduction of iron from its red-yellow ferric form to its blue-grey ferrous state.

Gravitational water Water in excess of *hygroscopic* and *capillary water*; transitory in soil and normally quickly drained by gravity.

Habitat Specific environment where plants and animals live; the 'home' of an organism or group of organisms.

Hard pan Hardened soil horizon caused by cementation of soil particles with organic matter, sesquioxides (e.g. iron pan), silica or calcium carbonate.

Horizon A layer of soil roughly parallel to the ground surface but differing from adjoining layers in physical, chemical and biological properties.

Humus Decomposed or partially decomposed plant and animal remains found in O-horizons and, in association with mineral matter, in A-horizons.

Hydrolysis The main type of chemical weathering in which water combines with rock minerals to form clay.

Hygroscopic water Water held in thin films close to the surface of soil particles and not available to plants.

Illuviation Process of deposition or washing in of soil material removed from one horizon to another; usually upper to lower in profile. Compare *eluviation*.

Intrazonal soil Soil whose characteristics are strongly modified by local factors, e.g. relief (landform), parent material, age, rather than by climate or vegetation.

Ion An atom or group of atoms with an electric charge.

Kaolin Highly weathered clay found in, but not exclusive to, tropical soils.

Leaching Specifically, the removal of base cations from the soil by acid rainwater. Sometimes used to denote the removal from the soil of all solutes (e.g. bases, dissolved clay, soluble organic matter, iron compounds) by acid drainage waters.

Mantle The uppermost layer of the earth's surface.

Micelle Individual clay particle.

Microclimate Climate of a small area within a few metres of the ground surface.

Mottles Spots of different colour in the soil. Often used in relation to patches of red ferric iron oxide within a blue-grey matrix or horizon of ferrous iron oxides.

Palaeosol A ancient soil horizon or buried fossil soil. Ideally should contain organic remains to allow estimation of age.

Parent material The principal source of weathered rock (*regolith*) from which the soil is formed.

Pedon Smallest volume which can be called 'a soil'; forms three-dimensional column. Similar pedons grouped together form *polypedons*.

Peri-glacial Of the climate, processes and features created by alternate freezing and thawing in a zone bordering ice sheets.

pH level See *acidity*.

Plagioclimax Plant community checked in its progress towards the *climatic climax* by human activity, e.g. heather moorlands in the UK.

Plant association A type of vegetation (plant community) characterised by a particular assemblage or collection of species; usually denoted by the dominant species, e.g. oak association.

Plant community Strictly, a distinctive collection of plants composed of two or more species and often related to particular environmental conditions, e.g. heathland community. Used loosely however to indicate any vegetation type.

Plant formation A plant community defined on the basis of its physical form or structure. Large plant formations related to climate are synonymous with *biomes*.

Pore (*soil*) Void or space between the soil particles. *Micropores* are tiny pores (e.g. between clay particles) and *macropores* are large pores (e.g. between sand grains).

Productivity (*plant*) The growth rate of vegetation, expressed as mass (e.g. dry weight of tissue) per unit area per unit time.

Rain-shadow The area on the leeward side of a mountain range where precipitation amounts are smaller than those on the rain-bearing, windward side.

Reaction (*soil*) Degree of acidity or alkalinity usually expressed in terms of pH value.

Regolith The weathered surface rocks or *mantle*.

Salinisation Process whereby salts are drawn to the surface of the soil and deposited by evaporation.

Sesquioxide An oxide of a trivalent element. In soils this refers to oxides of iron and aluminium.

Soil moisture deficit The amount by which soil moisture falls below *field capacity*.

Soil profile A vertical section of soil in which all the *horizons* are shown.

Soil structure Arrangement of individual (primary) soil particles into compound particles or aggregates.

Soil texture Relative proportion of individually sized mineral particles in the soil, e.g. sand, silt and clay.

Sub-climax vegetation Vegetation arrested in its development towards the *climatic climax* by various factors, e.g. *edaphic climax* is the result of local soil conditions (see *plagioclimax*).

Succession The sequence of vegetation change in a given site, when *plant communities* replace one another as they progress towards a climax stage. *Primary successions* are initiated on fresh, non-biologically modified sites; *secondary successions* are the result of disturbance and take place on previously vegetated sites.

Translocation A general term for the transfer of materials in soils.

Transpiration The loss by plants of water from the soil to the atmosphere.

Trophic level The level at which energy, in the form of food, is transferred from one group of organisms to another in the *food chain*; i.e. feeding level.

Water budget (*soil*) The amount of water available for drainage and run off. It is the difference between precipitation (rainfall) and *evapotranspiration*.

Zonal soils Major group of soils whose properties are dominated by the influence of climate and vegetation. At the global level, zonal soils, climax vegetation and global climatic types are closely associated.

References and Further Reading

Chapter 1

Agricultural Advisory Council (1970) *Modern Farming and the Soil*, Ministry of Agriculture, Fisheries and Food, HMSO, London.

Birkeland, P. W. (1984) *Soils and Geomorphology*, Oxford University Press, Oxford.

Brady, N. C. (1984) *The Nature and Properties of Soils*, Collier, Macmillan, New York.

Bridges, E. M. (1978) Soil, the vital skin of the earth, *Geography*, Vol. 63(4), pp. 354–61.

Courtney, F. M. and Trudgill, S. T. (1984) *The Soil: An Introduction to Soil Study*, Arnold, pp. 1–31.

Low, A. J. (1972) The effect of cultivation on the structure and other physical characteristics of grassland and arable soils, 1945–70, *Journal of Soil Science*, Vol. 23, pp. 363–80.

Munton, R. (1983) Agriculture and conservation: what room for compromise? In A. Warren and F. B. Goldsmith (eds.), *Conservation in Perspective*, Wiley, New York, pp. 353–73.

Newson, M. D. and Hanwell, J. D. (1982) *Systematic Physical Geography*, Macmillan, London, pp. 108, 114.

Shoard, M. (1980) *The Theft of the Countryside*, Temple Smith, London.

Simpson, K. (1983) *Soil*, Longman, London.

Soil Survey of England and Wales (1984) *Regional Bulletins* to accompany 1:250 000 Map of England and Wales, Soil Survey, Harpenden.

Tivy, J. (1982) *Biogeography: a Study of Plants in the Ecosphere*, Longman, London.

Warren, A. and Goldsmith, F. B. (1983) *Conservation in Perspective*, Wiley, New York.

Chapter 2

Birkeland, P. W. (1984) *Soils and Geomorphology*, Oxford University Press, Oxford, pp. 168, 241.

Burnham, C. P. (1970) The regional pattern of soil formation in Great Britain, *Scottish Geographical Magazine*, Vol. 86, pp. 25–34.

Burnham, C. P. (1980) The Soils of England and Wales, *Field Studies*, Vol. 5, pp. 349–63, with an Appendix and 1:2 000 000 Map by B. W. Avery, D. C. Findlay and D. Mackney.

Curtis, L. F. and Trudgill, S. T. (1976) *Soils in the British Isles*, Longman, London.

Etherington, J. R. (1982) *Environment and Plant Ecology*, Wiley, New York, pp. 39–82.

FitzPatrick, E. A. (1983) *Soils: Their Formation, Classification and Distribution*, Longman, London.

Pitty, A. F. (1979) *Geography and Soil Properties*, Methuen, London.

Simonson, R. W. (1978) A multiple-process model of soil genesis. In W. C. Mahaney (ed.), *Quaternary Soils*, Geo Abstracts, Norwich, pp. 1–25.

Smith, R. T. (1984) Soils in Ecosystems. In J. A. Taylor (ed.), *Themes in Biogeography*, Croom Helm, London, pp. 191–233.

Soil Survey of England and Wales (1984) *Soils and Their Use in Wales*, Bulletin Number 11, Soil Survey Harpenden.

Tivy, J. and O'Hare, G. (1981) *Human Impact on the Ecosystem*, Oliver and Boyd, Edinburgh, p. 18.

Trudgill, S. T. (1977) *Soil and Vegetation Systems*, Clarendon Press, Oxford.

Chapter 3

Allaby, M. (1983) *The Changing Uplands*, Countryside Commission, Cheltenham.

Cousens, J. (1974) *An Introduction to Woodland Ecology*, Oliver and Boyd, Edinburgh, pp. 10, 27.

Day, D. (1986) The Vegetation of the Sand Dune System at Gibraltar Point, Skegness. Undergraduate thesis, Derbyshire College of Higher Education, Derby.

Eyre, S. R. (1968) *Vegetation and Soils: A World Picture*, Arnold, p. 14.

Fitter, R. and Fitter, A. (1984) *Collins Guide to the Grasses, Sedges, Rushes and Ferns of Britain and Northern Europe*, Collins, Glasgow and London.

Gilbertson, D. *et al* (1985) *Practical Ecology for Geography and Biology: Survey, Mapping and Data Analysis*, Hutchinson, London.

Gimmingham, C. H. (1975) *An Introduction to Heathland Ecology*, Oliver and Boyd, Edinburgh.

Institute of Terrestrial Ecology (1978) *Upland Landuse in England and Wales*, Countryside Commission, Cheltenham, p. 87.

Institute of Terrestrial Ecology (1982) *Vegetation Change in Upland Landscapes*, Natural Environment Research Council, Swindon, pp. 12, 19.

Keble-Martin, W. (1982) *New Concise British Flora*, Michael Joseph, London.

Miles, J. (1981) *Effect of Birch on Moorlands*, Institute of Terrestrial Ecology, Natural Environment Research Council, Swindon, pp. 10, 11.

Rose, F. (1981) *Wild Flower Key: British Isles and North West Europe*, Warne, London.

Shimwell, D. W. (1984) Vegetation Analysis. In J. A. Taylor (ed.), *Themes in Biogeography*, Croom Helm, London, pp. 132–62.

Tivy, J. (1982) *Biogeography: a Study of Plants in the Ecosphere*, Longman, London, pp. 168–234.

Walker, J. *et al* (1981) Plant succession and soil development in coastal sand dunes of subtropical Eastern Australia. In D. C. West *et al* (eds.), *Forest Succession: Concepts and Application*, Springer-Verlag, New York, pp. 107–31.

Whittaker, R. H. (1975) *Communities and Ecosystems*, Macmillan, London, p. 175.

Chapter 4

Gersmehl, P. J. (1976) An alternative biogeography, *Annals of the Association of American Geographers*, Vol. 66, pp. 223–41.

Gosz, J. R. *et al* (1978) The flow of energy in a forest ecosystem, *Scientific American*, Vol. 238, pp. 93–102.

Hartzell, J. (1986) *Vanishing Earth*, Viewer's Guide to the B.B.C./T.V.E. co-production, International Broadcasting Trust.

Southwick, C. H. (1976) *Ecology and the Quality of Our Environment*, Van Nostrand, New York.

Tivy, J. and O'Hare, G. (1981) *Human Impact on the Ecosystem*, Oliver and Boyd, Edinburgh, p. 115.

Walter, H. (1973) *Vegetation on the Earth*, Springer-Verlag, New York, p. 13.

Chapter 5

Bridges, E. M. (1970) *World Soils*, Cambridge University Press.

Eyre, S. R. (1968) *Vegetation and Soils: a World Picture*, Arnold.

Furley, P. A. (1974) Soil-slope-plant relationships in the northern Maya Mountains, Belize, Central America, *Journal of Biogeography*, Vol. 1, pp. 263–79.

Furley, P. A. and Newey, W. W. (1982) *Geography of the Biosphere: an Introduction to the Nature, Distribution and Evolution of the World's Life Zones*, Butterworth, London, p. 311.

Hammond, A. L. (1972) Ecosystem analysis: biome approach to environmental research, *Science*, Vol. 175, pp. 46–48.

Hunt, C. B. (1966) *Plant Ecology of Death Valley, California*, U.S. Geological Survey, Professional Paper 509, San Francisco.

Hunt, C. B. (1972) *Geology of Soils*, Freeman, San Francisco, pp. 96, 106, 116, 126, 172.

Mabberley, D. J. (1983) *Tropical Rain Forest Ecology*, Blackie, Glasgow.

O'Hare, G. and Sweeney, J. (1986) *The Atmospheric System*, Oliver and Boyd, Edinburgh.

Pianka, E. R. (1982) *Evolutionary Ecology*, Harper and Row, New York.

Richards, P. W. (1979) *The Tropical Rain Forest*, Cambridge University Press.

Simmons, I. G. (1982) *Biogeographical Processes*, Allen and Unwin, London.

Tivy, J. (1982) *Biogeography: a Study of Plants in the Ecosphere*, Longman, London, pp. 258–376.

Walter, H. (1973) *Vegetation of the Earth*, Springer-Verlag, New York, pp. 20, 22, 38, 62, 116, 147, 210.

Webber, P. J. (1978) Spatial and temporal variation of the vegetation and its production, Barrow, Alaska. In L. L. Tieszen (ed.), *Vegetation and Production Ecology of an Alaskan Arctic Tundra*, Springer-Verlag, New York, pp. 37–112.

Chapter 6

Barke, M. and O'Hare, G. (1984) *The Third World: Diversity, Change and Interdependence*, Oliver and Boyd, Edinburgh, p. 42.

Duvigneaud, P. and Denaeyer-De Smet, S. (1970) Nutrient Cycling. In D. E. Reichle (ed.), *Analysis of Temperate Forest Ecosystems*, Chapman and Hall, pp. 199–225.

Furley, P. A. and Newey, W. W. (1982) *Geography of the Biosphere: An Introduction to the Nature, Distribution and Evolution of the World's Life Zones*, Butterworth, London.

Gersmehl, P. J. (1976) An alternative biogeography, *Annals of the Association of American Geographers*, Vol. 66, pp. 223–41.

Goudie, A. (1981) *The Human Impact: Man's Role in Environmental Change*, Blackwell, London, pp. 25–139.

Hartzell, J. (1986) *Vanishing Earth*, Viewer's Guide to the B.B.C./T.V.E. co-production, International Broadcasting Trust.

Seymour, J. and Giradet, H. (1986) *Far From Paradise: the Story of Man's Impact on the Environment*, B.B.C., London.

Simmons, I. G. (1982) *Biogeographical Processes*, Allen and Unwin, London.

Tivy, J. and O'Hare, G. (1981) *Human Impact on the Ecosystem*, Oliver and Boyd, Edinburgh.

Whittaker, R. H. (1975) *Communities and Ecosystems*, Macmillan, London, p. 224.

Chapter 7

Brown, L. and Wolf, E. C. (1984) *Soil Erosion: Quiet Crisis in the World Economy*, Worldwatch Paper No. 60, Worldwatch Institute, Washington D.C.

Bunyard, P. (1986) The death of the trees, *The Ecologist*, Vol. 16, pp. 4–13.

E.E.C. Directorate-General for Environment/Environmental Resources Ltd (1983) *Acid Rain: a Review of the Phenomenon in the E.E.C. and Europe*, Graham and Trotman, London, p. 39.

Hartzell, J. (1986) *Vanishing Earth*, Viewer's Guide to the B.B.C./T.V.E. co-production, International Broadcasting Trust.

Institute of Terrestrial Ecology (1984) *Agriculture and Environment*, Natural Environment Research Council, Swindon.

International Union for Conservation of Nature and Natural Resources *Red Data Books*, Gland, Switzerland.

Jackman, B. and Paskell, T. (1986) Battle for a natural Britain, *Sunday Times Magazine*, 20 April, 1986, pp. 56–59.

Kirkby, M. J. and Morgan, R. P. C. (1980) *Soil Erosion*, Wiley, New York.

Lean, G. (1986) Acid rain, *Observer Magazine*, 19 October, 1986, pp. 50–60.

Likens, G. E. *et al.* (1979) Acid rain, *Scientific American*, Vol. 241, pp. 39–47.

McCormick, J. (1985) *Acid Earth: the Global Threat of Acid Pollution*, Earthscan, International Institute for Environment and Development, London, p. 14.

Morgan, R. P. C. (1986) Soil erosion in Britain: the loss of a resource, *The Ecologist*, Vol. 16, pp. 40–41.

Nature Conservancy Council (1981) *Nature Conservation and Agriculture*, Nature Conservancy Council, p. 17.

Pye-Smith, C. and Rose, C. (1984) *Crisis and Conservation: Conflict in the British Countryside*, Penguin, London, pp. 2, 97.

Seymour, J. and Giradet, H. (1986) *Far From Paradise: the Story of Man's Impact on the Environment*, B.B.C., London.

Shoard, M. (1980) *The Theft of the Countryside*, Temple Smith, London.

Tivy, J. (1975) Environmental impact of cultivation. In J. M. A. and W. W. Fletcher (eds.), *Food, Agriculture and Environment*, Environment and Man, Vol. 2, Blackie, Glasgow, pp. 21–47.

Chapter 8

Briggs, D. (1977) *Soils*, Butterworth, London.

Trudgill, S. T. (1982) *Weathering and Erosion*, Butterworth, London.

Chapter 9

Avery, B. W. (1980) Soil Classification for England and Wales: Soil Survey Technical Monograph, No. 14, Soil Survey, Harpenden.

Bridges, E. M. and Davidson, D. A. (1982) *Principles and Applications of Soil Geography*, Longman, London, pp. 28–96.

Clarke, G. R. (1971) *The Study of Soil in the Field*, Oxford University Press.

Gregory, S. (1976) *Statistical Methods and the Geographer*, Longman, London.

Hodgson, J. M. (1978) *Soil Sampling and Soil Description*, Oxford University Press.

Jarvis, R. A. (1985) Bogged down in soil studies, *Teaching Geography*, June, 1985.

Soil Survey of England and Wales *Regional Bulletins* (1984) and *Legend* (1983) to accompany 1:250 000 Map of England and Wales, (6 sheets), Soil Survey, Harpenden.

Soil Survey of Scotland *Regional Handbooks* (1983) to accompany 1:250 000 Map of Scotland, (7 sheets), Soil Survey, Craigiebuckler, Aberdeen.

Index

mineral matter 8, 9, 10, 14, 24
moder 15, 41
monoculture 22, 155
mor humus 15, 32, 39, 41
mull humus 14, 32, 33, 37, 39, 46, 125
mycorrhizae 141

Nature Conservancy Council 154, 157
nature reserve 157
nitrogen 14, 20, 22, 102
nitrogen oxides 18, 158, 159, 162
nutrient
  cycling 23, 100–2, 140–1
  deficiency 13, 15, 22
  plant 10, 12, 14, 16, 17, 19, 24, 169

organic acids 18, 35, 41
organic matter 8, 12, 14, 15, 19, 31–3, 177, 192
oxidation 31, 120
ozone 157, 158

palaeosol 53, 54
parent materials 7, 9, 15, 31, 48–9, 189
peat 15, 32, 43–4
pedon 8
permafrost 111
pesticides 22, 23
pH (see acidity)
photosynthesis 97, 138
phreatophyte 130
pioneer species 65, 70
Pleistocene glaciation 53, 54
plant (vegetation)
  association 61
  community 60–2, 85
  distribution 64–5, 68
  dominance 88
  formation 62
  production 66, 97, 98, 138–9, 143
  stratification 62–4, 66, 85, 112, 113, 117
  survey 85–9

plinthite 118
plough pan 24
podzols 39–42, 44, 48, 49, 52, 53, 54, 112, 131, 145, 160
podzolisation 35, 41
pore (soil) 11, 16, 24, 35
post glacial period 54
prairie
  grassland 123
  soil 125
precipitation/evapotranspiration ratio 34, 35, 38, 43, 81, 124
primary minerals 9, 31

quadrat analysis 86–8

raw soils (lithosols) 45, 46, 51, 52, 131
raised bogs 43
reduction 31, 120
red yellow soil (podzol) 119
regolith 7, 54
rendzina 46
respiration 98
roots 13, 14, 16, 17, 192–3

salinisation 129, 143–5, 150
salt pan 128, 129
sampling 86
savanna 126, 147
secondary minerals 9, 31
sere 66
sesquioxides 10, 35
silicate clay 10, 31
Site of Special Scientic Interest (SSSI) 154, 157
soil
  air-dry fine earth fraction 176, 177
  analyses 176–80
  as open system 30–1
  bulk density 24, 178
  colour 120, 191
  depth 19, 31, 45, 48, 52
  forming factors 29, 46
  horizons 7, 14, 29, 32, 37, 190, 191
  mottles 192
  pit 185, 186

soil (cont.)
  profile 7, 29, 174, 186
  solution 10, 17, 18
  structure 9, 15, 16, 19, 20, 23, 24, 124, 167, 191–2
  Survey (National) 170, 183
  texture 9, 10–13, 49, 178–80
  throughflow 33, 51
  type 8, 29, 37
species number (diversity) 60, 66, 79, 111, 113, 117, 131, 154
steady state soil (see equilibrium soil)
succession (plant) 65–85
  in lakes 70–1
  on sand dunes 71–9
  primary 67, 69–79
  retrogressive 68
  secondary 67, 80–5
sulphur dioxide 18, 158–62

taiga (see coniferous forest)
temperate deciduous forest 62, 63, 139, 140–1
terracing 165, 168
thermokarst 142
time factor 52–4, 65
topsoil 7, 14, 15
trace element 17, 23
translocation 35, 125
transpiration 12, 33, 122
trophic level 98–9
tropical rain forest 60, 68, 95, 116–22, 140–1, 146–7
tundra 68, 95, 130–2, 139, 142

vegetation (see plant)

water budget 33–5
waterlogging 13, 15, 24, 34, 35, 40, 43, 49, 51, 167
weathering 9, 14, 17, 31, 45, 48, 52, 101
wilting point 12

xerophytes 130

zonal soils 68, 108